T0328345

DATA VISUALIZATION AND ANALYSIS IN SECOND LANGUAGE RESEARCH

This introduction to visualization techniques and statistical models for second language research focuses on three types of data (continuous, binary, and scalar), helping readers to understand regression models fully and to apply them in their work. Garcia offers advanced coverage of Bayesian analysis, simulated data, exercises, implementable script code, and practical guidance on the latest R software packages. The book, also demonstrating the benefits to the L2 field of this type of statistical work, is a resource for graduate students and researchers in second language acquisition, applied linguistics, and corpus linguistics who are interested in quantitative data analysis.

Guilherme D. Garcia is Assistant Professor of Linguistics at Ball State University, USA.

SECOND LANGUAGE ACQUISITION RESEARCH SERIES

Susan M. Gass and Alison Mackey, Series Editors
Kimberly L. Geeslin, Associate Editor

The *Second Language Acquisition Research Series* presents and explores issues bearing directly on theory construction and/or research methods in the study of second language acquisition. Its titles (both authored and edited volumes) provide thorough and timely overviews of high-interest topics and include key discussions of existing research findings and their implications. A special emphasis of the series is reflected in the volumes dealing with specific data collection methods or instruments. Each of these volumes addresses the kinds of research questions for which the method/instrument is best suited, offers extended description of its use, and outlines the problems associated with its use. The volumes in this series will be invaluable to students and scholars alike and perfect for use in courses on research methodology and in individual research.

Eye Tracking in Second Language Acquisition and Bilingualism
A Research Synthesis and Methodological Guide
Aline Godfroid

Theories in Second Language Acquisition
An Introduction, Third Edition
Edited by Bill VanPatten, Gregory D. Keating and Stefanie Wulff

Data Visualization and Analysis in Second Language Research
Guilherme D. Garcia

For more information about this series, please visit: www.routledge.com/Second-Language-Acquisition-Research-Series/book-series/LEASLARS

DATA VISUALIZATION AND ANALYSIS IN SECOND LANGUAGE RESEARCH

Guilherme D. Garcia

NEW YORK AND LONDON

First published 2021
by Routledge
605 Third Avenue, New York, NY 10158

and by Routledge
2 Park Square, Milton Park, Abingdon, Oxon, OX14 4RN

Routledge is an imprint of the Taylor & Francis Group, an informa business

Library of Congress Cataloging-in-Publication Data
A catalog record for this book has been requested

ISBN: 978-0-367-46964-1 (hbk)
ISBN: 978-0-367-46961-0 (pbk)
ISBN: 978-1-003-03224-3 (ebk)

Typeset in Bembo
by Apex CoVantage, LLC

CONTENTS

List of Figures x
List of Tables xii
List of Code Blocks xiii
Acknowledgments xvi
Preface xviii

Part I
Getting Ready **1**

1 Introduction 3
1.1 Main Objectives of This Book 3
1.2 A Logical Series of Steps 5
1.2.1 Why Focus on Data Visualization Techniques? 5
1.2.2 Why Focus on Full-Fledged Statistical Models? 6
1.3 Statistical Concepts 7
1.3.1 p-*Values 7*
1.3.2 Effect Sizes 9
1.3.3 Confidence Intervals 10
1.3.4 Standard Errors 11
1.3.5 Further Reading 12

2 R Basics 14
2.1 Why R? 14
2.2 Fundamentals 16

2.2.1 *Installing R and RStudio* 16
2.2.2 *Interface* 17
2.2.3 *R Basics* 20
2.3 *Data Frames* 28
2.4 *Reading Your Data* 31
2.4.1 *Is Your Data File Ready?* 32
2.4.2 *R Projects* 32
2.4.3 *Importing Your Data* 33
2.5 *The* Tidyverse *Package* 35
2.5.1 *Wide-to-Long Transformation* 36
2.5.2 *Grouping, Filtering, Changing, and Summarizing Data* 39
2.6 *Figures* 42
2.6.1 *Using* Ggplot2 43
2.6.2 *General Guidelines for Data Visualization* 46
2.7 *Basic Statistics in R* 49
2.7.1 *What's Your Research Question?* 50
2.7.2 t-*Tests and ANOVAs in R* 51
2.7.3 *A* Post-Hoc *Test in R* 53
2.8 *More Packages* 55
2.9 *Additional Readings on R* 55
2.10 *Summary* 55
2.11 *Exercises* 57

Part II
Visualizing the Data **61**

3 Continuous Data 63
3.1 *Importing Your Data* 65
3.2 *Preparing Your Data* 66
3.3 *Histograms* 68
3.4 *Scatter Plots* 70
3.5 *Box Plots* 75
3.6 *Bar Plots and Error Bars* 77
3.7 *Line Plots* 80
3.8 *Additional Readings on Data Visualization* 82
3.9 *Summary* 82
3.10 *Exercises* 83

4 Categorical Data 86
4.1 *Binary Data* 88

4.2 Ordinal Data 94
4.3 Summary 97
4.4 Exercises 97

5 Aesthetics: Optimizing Your Figures 99
5.1 More on Aesthetics 104
5.2 Exercises 104

Part III
Analyzing the Data 107

6 Linear Regression 109
6.1 Introduction 111
6.2 Examples and Interpretation 115
6.2.1 Does Hours *Affect Scores? 115*
6.2.2 Does Feedback *Affect Scores? 119*
6.2.3 Do Feedback *and* Hours *Affect Scores? 123*
6.2.4 Do Feedback *and* Hours *Interact? 126*
6.3 Beyond the Basics 131
6.3.1 Comparing Models and Plotting Estimates 131
6.3.2 Scaling Variables 134
6.4 Summary 139
6.5 Exercises 140

7 Logistic Regression 143
7.1 Introduction 144
7.1.1 Defining the Best Curve in a Logistic Model 148
7.1.2 A Family of Models 148
7.2 Examples and Interpretation 149
7.2.1 Can Reaction Time Differentiate Learners and Native Speakers? 150
7.2.2 Does Condition *Affect Responses? 156*
7.2.3 Do Proficiency *and* Condition *Affect Responses? 159*
7.2.4 Do Proficiency *and* Condition *Interact? 163*
7.3 Summary 170
7.4 Exercises 171

8 Ordinal Regression 173
8.1 Introduction 174
8.2 Examples and Interpretation 175

8.2.1 Does Condition Affect Participants' Certainty? 175
8.2.2 Do Condition and L1 Interact? 181
8.3 Summary 185
8.4 Exercises 187

9 Hierarchical Models 189
9.1 Introduction 189
9.2 Examples and Interpretation 194
9.2.1 Random-Intercept Model 195
9.2.2 Random-Slope and Random-Intercept Model 198
9.3 Additional Readings on Regression Models 205
9.4 Summary 207
9.5 Exercises 207

10 Going Bayesian 210
10.1 Introduction to Bayesian Data Analysis 212
10.1.1 Sampling From the Posterior 217
10.2 The RData Format 221
10.3 Getting Ready 222
10.4 Bayesian Models: Linear and Logistic Examples 223
10.4.1 Bayesian Model A: Feedback 223
10.4.2 Bayesian Model B: Relative Clauses with Prior Specifications 229
10.5 Additional Readings on Bayesian Inference 233
10.6 Summary 235
10.7 Exercises 236

11 Final Remarks 240
Appendix A: Troubleshooting 242
A.1 Versions of R and RStudio 242
A.2 Different Packages, Same Function Names 242
A.3 Errors 243
A.4 Warnings 244
A.5 Plots 244
Appendix B: RStudio Shortcuts 245
Appendix C: Symbols and Acronyms 246
Appendix D: Files Used in This Book 247
Appendix E: Contrast Coding 250
Appendix F: Models and Nested Data 251

Glossary 253
References 257
Subject Index 261
Function Index 263

FIGURES

2.1	RStudio Interface and Pane Layout (Mac Version)	18
2.2	Starting Your First Script with RStudio: Math Operations	23
2.3	Bar Plot with Standard Error Bars Using ggplot2	43
2.4	Three Variables in a Bar Plot Using ggplot2	48
3.1	Organizing Your Files with Two R Projects	65
3.2	A Typical Histogram Using ggplot2	69
3.3	A Typical Scatter Plot Using ggplot2	70
3.4	A Scatter Plot with Multiple Trend Lines Using ggplot2	72
3.5	A Scatter Plot with Point Size Representing a Continuous Variable	74
3.6	The Structure of a Box Plot	75
3.7	A Box Plot (with Data Points) Using ggplot2	76
3.8	A Bar Plot (with Error Bars) Using ggplot2	78
3.9	A Box Plot with Error Bars for Standard Errors and ±2 Standard Deviations	80
3.10	A Line Plot Using ggplot2	81
4.1	Resulting Plot after Data Transformation	89
4.2	A Visual Representation of props: participant s1	92
4.3	By-Speaker Responses	93
4.4	Plotting Ordinal Data with Bars	95
5.1	Plotting Ordinal Data with Adjustments	100
6.1	Fitting a Line to the Data	111
6.2	Participants' Scores by Weekly Hours of Study	116
6.3	Participants' Scores Across Both Feedback Groups	120

6.4 Participants' Scores by Weekly Hours of Study and by Feedback Type 124
6.5 A Figure Showing How Feedback and Hours May Interact 126
6.6 Plotting Estimates from a Linear Model 132
6.7 Rescaling a Binary Variable: Feedback.std 137
7.1 Native Language As a Function of Reaction Time: L1 ~ RT 144
7.2 Modeling a Binary Response As a Function of a Continuous Predictor 145
7.3 Odds and Log-odds 147
7.4 Predicted Probabilities for Ten Reaction Times (Using Model Fit) 153
7.5 Preference for Low Attachment by Condition 156
7.6 Preference for Low Attachment by Condition and Proficiency 160
7.7 Plotting Estimates from a Logistic Model 168
8.1 A Typical Scale with Labeled End-Points 174
8.2 Certainty Levels (%) by Condition for English and Spanish Groups 176
8.3 Calculating Probabilities in Ordinal Models (Condition = NoBreak) 180
8.4 Predicted Probabilities Using ggplot2 186
9.1 Grouped Data in longFeedback 190
9.2 Simple Model (a) *vs.* Hierarchical Models (b–c) 192
9.3 Effect of Feedback and Task on Participants' Scores 195
9.4 Shrinkage from Hierarchical Model 202
9.5 Plotting Estimates for Fixed and Random (by-Participant) Effects 204
10.1 Bayes's Rule in Action: Prior, Likelihood, and Posterior Distributions 213
10.2 How a Strong Prior Affects our Posterior 215
10.3 How a Larger Sample Size Affects our Posterior 215
10.4 95% Highest Density Interval of Posterior From Fig. 10.1 217
10.5 Bayesian Models in R Using Stan 218
10.6 Illustrative Posterior Distributions for $\hat{\beta}_0$ and $\hat{\beta}_1$ 220
10.7 Trace Plot for Diagnostics in Bayesian Model 226
10.8 Plotting Model's Estimates (Posterior Distributions) 228
10.9 Posterior Predictive Check 228
10.10 Posterior Distributions for All Parameters of Interest (Four Chains Each) 232
10.11 Posterior Distributions for Two Parameters 233
D.1 File Structure Used in This Book 248

TABLES

2.1 A Table in Wide Format 37
2.2 A Table in Long Format 37
2.3 Main Functions in dplyr 40
3.1 Excerpt of feedback: Wide Format (60 by 17) 66
3.2 Excerpt of feedback: Long (Tidy) Format (600 by 10) 68
6.1 Example of Contrast Coding 121
6.2 Statistical Table Showing Model Estimates 130
7.1 Probability (*P*), Odds, and ln(odds) 146
7.2 Model Estimates and Associated Standard Errors, *z*-values, and *p*-values 167
9.1 Model Specification in R Using Task as a Predictor 193
9.2 Model Estimates and Associated SEs 203
B.1 Useful Keyboard Shortcuts in RStudio 245
C.1 Main Symbols and Acronyms Used in the Book 246
D.1 List of all Scripts and their Respective Code Blocks 249
E.1 Example of Contrast Coding for More Than Two Levels 250

CODE BLOCKS

1 Variables and Vectors in R 26
2 Lists in R 27
3 Creating a Data Frame in R 30
4 Importing Your Data into R 35
5 Sourcing Scripts and Installing and Loading Packages in R 36
6 Wide-to-Long and Long-to-Wide Transformations Using tidyverse 39
7 Data Frame before and after Wide-to-Long Transformation Using tidyverse 40
8 Grouping, Summarizing, and Sorting Data with tidyverse 41
9 Generating Your First Plot Using ggplot2 44
10 Running *t*-Tests and ANOVAs in R (Results Given As Comments) 54
11 Preparing our Data for Plotting (and Analysis) 68
12 Producing a Histogram Using ggplot2 70
13 Producing a Scatter Plot Using ggplot2 72
14 Producing a Scatter Plot with Multiple Trend Lines Using ggplot2 73
15 Producing a Scatter Plot with Three Continuous Variables Using ggplot2 74
16 Producing a Box Plot Using ggplot2 77
17 Producing a Bar Plot Using ggplot2 79
18 Adding Error Bars to a Box Plot 81
19 Producing a Line Plot Using ggplot2 82
20 Preparing Binary Data for Bar Plot with Error Bars 91

21 Adding by-Participant Lines to a Bar Plot 93
22 Preparing the Data and Plotting Certainty Levels 96
23 Plotting Certainty Levels with Adjustments: Preparing the Data 101
24 Plotting Certainty Levels with Adjustments: Creating Plot 102
25 Simple Linear Regression and Output with Estimates: Score ~ Hours 117
26 Simple Linear Regression and Output with Estimates: Score ~ Feedback 122
27 Multiple Linear Regression: Score ~ Feedback + Hours 125
28 Modeling an Interaction: Score ~ Feedback * Hours 128
29 Comparing Models Using anova() 131
30 Preparing the Data for Plotting Model Estimates and Confidence Intervals 133
31 Plotting Model Estimates and Confidence Intervals 134
32 Rescaling Variables in a Linear Model 136
33 Preparing our Data for Logistic Models 150
34 Running a Logistic Regression in R 151
35 Assessing Model's Accuracy: fit glm1 154
36 Modeling Responses by Condition 158
37 Code for Fig. 7.6: Bar Plot and Error Bars (Three Variables) 161
38 Running a Logistic Regression with Two Categorical Predictors 162
39 Running a Logistic Regression with an Interaction 165
40 Creating a Plot for Model Estimates 169
41 Assessing Model's Accuracy: fit glm4 170
42 Plotting Certainty As a Function of L1 177
43 Modeling Certainty As a Function of Condition 178
44 Predicted Probabilities for Ordinal Model 182
45 Modeling Certainty As a Function of Condition and L1 183
46 How to Plot Predicted Probabilities for Ordinal Model 187
47 Updated Content of dataPrepLinearModels.R 194
48 Plotting Scores by Feedback and Task 195
49 Running a Model with Random Intercepts: Feedback and Task 196
50 Running a Model with Random Intercepts and Slopes: Feedback and Task 200
51 Random Effects in a Hierarchical Model (lmer) 201
52 Predicting Scores Using predict() 202
53 Plotting Fixed and Random Effects Manually (Part 1) 205
54 Plotting Fixed and Random Effects Manually (Part 2) 206

55 Fitting a Bayesian Hierarchical Linear Regression 224
56 Creating a Trace Plot 227
57 Code for Plotting Model's Estimates (Fig. 10.8) 228
58 Specifying Priors and Running our Model 231
59 Plotting Posterior Distributions (Figs. 10.10 and 10.11) 234
60 Running Multiple Models and Extracting Coefficients 252

ACKNOWLEDGMENTS

I wrote most of this book during the COVID-19 pandemic in 2020—in part thanks to a research grant I received from Ball State University (ASPiRE Junior Faculty Award). I was fortunate to be engaged in a project that not only interested me a lot but could be executed from home—and that kept me sufficiently busy during that stressful year. Thankfully, writing this book was perfectly compatible with the need for social distance during the pandemic.

My interest in quantitative data analysis concretely started during my first years as a graduate student at McGill University, back in 2012. During my PhD (2012–2017), I developed and taught different workshops involving data analysis and the R language in the Department of Linguistics at McGill as well as in the Department of Education at Concordia University. After joining Ball State in 2018, I also organized different workshops on data analysis. This book is ultimately the result of those workshops, the materials I developed for them, and the feedback I received from numerous attendees from different areas (linguistics, psychology, education, sociology, and others).

Different people played a key role in this project. Morgan Sonderegger was the person who first introduced me to R through graduate courses at McGill. Natália Brambatti Guzzo provided numerous comments on the book as a whole. This book also benefitted from the work of three of my students, who were my research assistants over the summer of 2020: Jacob Lauve, Emilie Schiess, and Evan Ward. They not only tested all the code blocks in the book but also had numerous suggestions on different aspects of the manuscript. Emilie and Jacob were especially helpful with the packaging of some sections and exercises. Different chapters of this book benefitted from helpful feedback from Jennifer Cabrelli, Ronaldo Lima Jr., Ubiratã Kickhöfel Alves, Morgan Sonderegger, Jiajia Su, Lydia White, and Jun Xu.

The chapter on Bayesian data analysis is an attempt to incentivize the use of Bayesian data analysis in the field of second language research (and in linguistics more generally). My interest in Bayesian statistics began in 2015, and a year later I took a course on Bayesian data analysis with John Kruschke at Universität St. Gallen. The conversations we had back then were quite helpful, and his materials certainly influenced my views on statistics and data visualization as a whole.

Finally, I wish to thank the team at Routledge, especially Ze'ev Sudry and Helena Parkinson, with whom I exchanged dozens of emails throughout 2019 and 2020. I'm grateful for their patience and work while copy-editing and formatting this book.

PREFACE

In this book, we will explore quantitative data analysis using visualization techniques and statistical models. More specifically, we will focus on *regression analysis*. The main goal here is to move away from *t*-tests and ANOVAs and towards full-fledged statistical models. Everything we do will be done using R (R Core Team 2020), a language developed for statistical computing and graphics. No background in R is necessary, as we will start from scratch in chapter 2 (we will also go over the necessary installation steps).

The book is divided into three parts, which follow a logical sequence of steps. In Part I, we will get you started with R. In Part II, we will explore data visualization techniques (starting with continuous data and then moving on to categorical data). This will help us understand what's going on in our data *before* we get into the statistical analysis. Finally, in Part III, we will analyze our data using different types of statistical models—Part III will also introduce you to Bayesian statistics. Much like Part II, Part III also covers continuous data before categorical data.

It's important to note that this is *not* a book about statistics *per se*, nor is it written by a statistician: I am a linguist who is interested in phonology and second language acquisition and who uses statistical methods to better understand linguistic data, structure, and theory. Simply put, then, this is a book about the *application* of statistical methods in the field of second language acquisition and second language research.

Intended Audience

This book is for graduate students in linguistics as well as second language acquisition researchers more generally, including faculty who wish to update their quantitative methods. Given that we will focus on data visualization and statistical models, qualitative methods and research design will not be discussed.

The data used in the book will be focused on topics which are specific to linguistics and second language research. Naturally, that doesn't mean your research must be aligned with second language acquisition for this book to be beneficial.

Background Knowledge

As mentioned earlier, this book does *not* expect you to be familiar with R. You are expected to be somewhat familiar with basic statistics (e.g., means, standard deviations, medians, sampling, *t*-tests and ANOVAs). We will briefly review the most crucial concepts in §1.3 later, and they will be discussed every now and then, but some familiarity is desirable. Bear in mind that the whole idea here is to move away from *t*-tests and ANOVAs towards more comprehensive, flexible, and up-to-date methods, so you should not worry too much about it if your memory is not so great when it comes to *t*-tests and ANOVAs.

How to Use This Book

Most chapters in this book are intended to be read in order. For example, to understand chapter 7 (logistic regressions), you should read chapter 6 (linear regressions)—unless you are already familiar with linear regressions.

Because we will use R throughout the book, you will be able to reproduce all the analyses and all the relevant figures *exactly*. This is possible because R is a language, and all the instructions to produce a figure, for example, are text-based, so they don't depend on mouse clicks, as in IBM's Statistical Package for the Social Sciences (SPSS).

R is a computer language, so you have to learn its vocabulary and its syntax (i.e., its "grammar"). To help you with that, this book contains numerous code blocks, which will allow you to reproduce the analyses presented in different chapters—make sure you carefully examine these blocks. You should manually type the code in R every time: it can be a little time-consuming, but it's certainly the best way to learn R—make sure you double-check your code, as it's very easy to miss a symbol or two. As you progress through the book, you will be able to adapt different code blocks to best suit your needs. Don't worry: we will go over code blocks line by line.

Different chapters also contain summaries with key points you should remember: short summaries for this preface and chapter 1 and expanded summaries for all the other chapters (except for chapter 6). Some chapters are accompanied by practice exercises. Finally, reading suggestions will be made in case you wish to learn more about specific topics covered in this book.

Many suggestions can be easily found online, and combining those resources with the present book will certainly strengthen your understanding of the topics we will explore.

As you implement the R code in this book, you might run into warning or error messages in R. Most of the time, warning messages are simply informing you about something that just happened, and you don't need to do anything. Error messages, on the other hand, will interrupt your code, so you will need to fix the problem before proceeding. Appendix A (Troubleshooting) can help with some of those messages. Alternatively, simply googling a message will help you figure out what's going on—the R community online is huge and very helpful. All the code in this book has been tested for errors on computers running Mac OS, Linux Ubuntu, and Windows 10, so you should not have any substantial issues.

Download the Necessary Files

To visualize and analyze data, we need data files. You can download all the necessary files from http://osf.io/hpt4g (Open Science Framework). The file you need is called `DVASLR-DATA.zip`. You can download it now if you want to, but we won't use this until chapter 2.

An Important Note on Organization

Throughout the book we will create multiple folders and files to keep things organized. I strongly recommend that you follow the file structure implemented here—especially if you're new to R and/or don't feel comfortable with file management. As we go through different chapters, we will create R files and add code blocks to them—I will tell you what to do and how to best organize your files whenever we come across a new file. Appendix D (*Files used in this book*) lists all the folders and files you will have created by the end of the book—including where each code block should be placed (Table D.1). I strongly recommend that you visit Appendix D every now and then. File organization is an important habit to develop from the beginning when it comes to data analysis. If your files are not organized, it's harder to find specific code blocks later on, and it's hard to understand what the code is doing.

Conventions, Terminology, and Language

Throughout this book, you will see some statistical symbols as well as a lot of words that use a slightly different font family. You will easily recognize these words because they use a `monospaced font`, so they will not look like typical words. These will be used to represent code in general, as it's easier to read code with such fonts. As a result, all the code blocks in the book will contain monospaced fonts as well. Words in **bold** represent terms that may not be familiar and which are explained in the glossary at the end of this book. Lastly, a list with the main statistical symbols and acronyms used in this book can be found in Appendix C.

When we discuss coding in general, we tend to use words and constructions which are not necessarily used in everyday language and which may be new to people who have never seen, studied, or read about code in general. For example, files that contain code are typically referred to as *scripts* (as mentioned earlier). Scripts contain several lines of code as well as comments that we may want to add to them (general notes). When we *run* a line of code we are essentially asking the computer to interpret the code and do or *return* whatever the code is requesting. You will often see the word *directory*, which simply means "folder where the relevant files are located".

In our scripts, we will almost always create **variables** that hold information for us (just like variables in math), so that later we can *call* these variables to access the information they hold. For example, when we type and run `city = "Tokyo"` in R, we create a variable that holds the value `Tokyo` in it. This value is simply a word in this case, which we can also refer to as a **string**. Later, if we just run `city`, we will get `Tokyo`, that is, its content, printed on the screen. When we import our data into R, we will always *assign* our data to a variable, which will allow us to reference and work on our dataset throughout our analysis.

A line of code is our way of asking R to do something for us: instead of clicking on buttons, we will write a line of code. For example, we may want to store information (e.g., `city`) or *evaluate* some expression (e.g., `5 + 3`). In the latter case, R will *print* the answer on the *console*—that is, it will display the answer in a specific area of your screen. If R finds a problem, it will *throw* an error message on the screen.

Lines of code throughout this book will involve *commands* or **functions**. Functions always have a name and brackets and are used to perform a task for us: for example, the function `print()` will print an object on the screen. The object we want to print needs to go inside the function as one of its *arguments*. Each function expects different types of arguments (they vary by function), and if we fail to include a required argument in a function, we will usually get an error (or a warning). This is all very abstract right now, but don't worry: we will explore all these terms in detail starting in chapter 2, which introduces the R language—and as you work your way through the book they will become familiar (and concrete).

Finally, all the important points discussed earlier can be found in the summary here. I hope that your own research can benefit from this book and that you enjoy learning how to visualize and analyze your data using R.

HOW TO USE THIS BOOK: SUMMARY

- An R file is called a *script*
- Throughout the book we will create several scripts
- Appendix D has the entire file structure adopted in this book
- Carefully work your way through the code blocks in each chapter
- Spend some time working on the exercises at the ends of chapters
- Use chapter 2 as your main reference for coding in R
- Consult Appendix A if you run into errors or warnings
- Consult Appendix C for a list of relevant symbols and acronyms used in this book
- Words such as "call", "run", "print", "object", "function", and "argument" have specific meanings when we're talking about code—you will get used to that as you read the chapters that follow

PART I

Getting Ready

1

INTRODUCTION

1.1 Main Objectives of This Book

It goes without saying that the field of second language acquisition (SLA), or second language research more generally, relies on data: test scores, reaction times, grammaticality judgements, certainty levels, categorical preferences, suppliance rates, and so on. Indeed, by the end of the 1990s, over 90% of all studies in applied linguistics were quantitative (Cunnings 2012). Different theories have been proposed to explain acquisition patterns that are observed in the data and to predict patterns in *unobserved* data. To evaluate the validity of any claim in the field, theoretical or not, we must consider empirical evidence, that is, *data*—much like any scientific field. It is therefore unsurprising that most of what we assume and propose in the field depends on *how* we analyze our data and, perhaps even more importantly, on how carefully we interpret and generalize the patterns that we find. More often than not, inappropriate analyses lead to incorrect conclusions.

It has been noted in the literature that the field of second language acquisition relies on a precariously narrow range of statistical techniques to quantitatively analyze data (Plonsky 2013, 2014, 2015). ***t*-tests** and **ANOVAs** still seem to be the most popular statistical options, even though they are (i) underpowered and (ii) often inappropriate given the data at hand, as will be discussed in Part III. As examples of (i), *t*-tests can't handle multiple variables (or groups), and ANOVAs can't handle complex hierarchical structures in the data (chapter 9), which are essential given how much variation we observe in linguistic data. In addition, both methods focus on *p*-values, not on effect sizes. An example of (ii) would include the use of ANOVAs when we are dealing with binary or scalar responses (chapters 7 and 8)—see Jaeger (2008).

Other problems in second language (L2) research include multiple tests in a single study, which drastically increase the probability of Type I error; emphasis on categorical and simplistic conclusions derived solely from *p*-values (i.e., a result is or isn't statistically significant, a common issue in null hypothesis significance testing); focus on a single variable, without considering the intricacies of language processes; and inappropriate data transformation (e.g., running ANOVAs on a categorical variable, as mentioned earlier, by first transforming it into percentages). These are just a small subset of many important methodological problems in our field.

The importance of reliable data analysis cannot be overstated, especially when replication studies are so rare in our field (see, e.g., Marsden et al. 2018. If numerous second language acquisition studies employ weak statistical methods, and almost no study is replicated, how worried should we be about the reliability of said studies? The fact that we have little access to published data and study designs doesn't help, naturally. As more and more data is made available through open science initiatives, however, we will be better equipped to evaluate studies that have played a central role in building the traditional assumptions that we typically take for granted. Crucially, then, improving our data analysis techniques helps us better evaluate what has been done in the field and also increases the reliability of what we do *now*.

The present book has two main goals that address data analysis in L2 research. The first goal is to provide better methods to explore patterns in data, especially methods for data visualization. The second goal is to replace commonly used statistical tests and ANOVAs with full-fledged statistical models. Not only are these models more flexible and powerful, they are more realistic given the complexity of real data patterns that we observe in L2 research. Ultimately, these goals have a common underlying ideal: *to provide a more refined statistical toolset, where we focus on evaluating the uncertainty of a given effect.*

To accomplish these two goals, we will rely on the R language (R Core Team 2020). The advantages of using R as opposed to using, for example, SPSS, are discussed in §2.1. Because the book assumes you have never used R, chapter 2 provides a brief introduction to the language, where you will learn everything you need to use this book without any other sources. Of course, other sources will always improve your understanding, so different reading suggestions will also be provided throughout the book, both on statistical concepts and on the R language itself.

As we shift our focus from *t*-tests (or other statistical tests) and ANOVAs to full-fledged statistical models, we will also shift the focus of the questions we will ask when examining our results. Among other topics, we will discuss whether our data meets the necessary assumptions behind a particular method; why we should emphasize effect sizes, not *p*-values; why confidence intervals matter—and why they are counterintuitive; and many other key concepts.

We will explore three types of models—yes, only three. They will handle three different types of data and will equip you with the most fundamental tools to analyze your data. Most of what we do in L2 research involves continuous data (e.g., scores, reaction time), binary data (e.g., yes/no, correct/incorrect), and ordinal data (e.g., Likert scales). By the end of this book, we will have examined models that can handle all three situations.

You may have noticed that almost all statistical analyses in our field rely on *p*-values. This is a central characteristic in Frequentist statistics (i.e., traditional statistical methods). But there is another possibility: Bayesian data analysis, where *p*-values do not exist. This book has a chapter dedicated to Bayesian statistics (chapter 10) and discusses why second language research can benefit from such methods.

This book assumes that you have taken some introductory course on research methods. Thus, concepts like mean, median, range, and *p*-values should sound familiar. You should also know a thing or two about ANOVAs and *t*-tests. However, if you think you may not remember these concepts well enough, §1.3 will review the important bits that will be relevant to us here. If you think you don't have enough experience running *t*-tests, that's OK: if this book is successful, you will not use *t*-tests anyway.

1.2 A Logical Series of Steps

The present book is based on a simple series of steps involved in quantitative data analysis. You can think of these steps as merely verbs that apply to your data: *import, manipulate, explore and visualize, analyze, interpret.* Every chapter in Part III will go through these steps, simply because they mirror what you would normally have to do with your own data—Part II will provide visualization techniques to prepare you for Part III. Some of these steps are obvious: you need to import your data before you can analyze it, of course. But some are not: not everyone explores data patterns with plots and summaries *before* running some statistical test (and some may not do that ever).

1.2.1 Why Focus on Data Visualization Techniques?

Good figures are crucial. When we want to see and understand the data patterns underlying a study, a figure is likely the best option we have. Whether or not you consider yourself a visual person, the truth is that a well-designed plot almost always communicates your results more effectively than huge tables or a series of numbers and percentages within the body of the text. An appropriate figure can help the reader understand your thinking process, your narrative, and your statistical results—needless to say, you *want* your reader (or your reviewer) to understand *exactly* what you mean.

Besides helping the reader, figures help *us*, the researchers. To design the right figure, we must have a clear understanding of the message we want to communicate, the pattern we want to focus on. Often times, it is by designing a figure that we realize that we don't know exactly what we want to see—or that we don't understand the type of data we are trying to visualize. In that way, data visualization can also improve our thinking process.

Despite the importance of data visualization, it is not uncommon to come across experimental papers in L2 research with very limited figures. How many times have you seen pie charts in papers? Or maybe bar plots without error bars? Perhaps a plot showing variables that do not match the variables discussed later, in the actual analysis? As we will see throughout this book, some general "rules of thumb" can already drastically improve one's data visualization techniques.

1.2.2 Why Focus on Full-Fledged Statistical Models?

Data can be messy, and language data almost always is—especially when we deal with L2 research, where multiple grammars can be at play, generating multiple patterns that often seem contradictory. Different learners behave differently, and they may respond differently to different stimuli or circumstances. On top of all that, we know that multiple variables can play a role in different phenomena. We naturally want to focus on a particular variable of interest, but we shouldn't ignore potential confounding factors that could also affect our results.

In the past, comparing two groups and concluding that they were statistically different was often good enough. Take *t*-tests, for example, which are usually used when we focus our attention on a single question: are two groups (i.e., means) statistically different? The answer is categorically defined based on an arbitrary number: 0.05 (see review on *p*-values in §1.3.1). This approach often doesn't care about *how* different the groups are—although this seems to be more and more a characteristic of the past now (i.e., what is generally referred to as *old statistics*). In contrast, the focus of statistical analyses these days has shifted a lot in many areas. We now typically focus on *models* with multiple variables because we wish to estimate the *effect* (§1.3.2) that such variables have on an outcome of interest.

Statistical models are very different from a simple *t*-test. They allow us to incorporate a high degree of complexity and therefore provide more realistic and reliable results. We can, for example, take into account how much learners will vary when it comes to the effect we are examining in our study. By default, such models also provide effect sizes, which are more relevant than mere *p*-values (cf. ANOVAs and *t*-tests, which require additional steps to yield effect sizes). They also allow us to measure the effect of multiple variables—not just one. These are only a few reasons that you should always favor a full-fledged statistical model over statistical tests—or simple ANOVAs, which are technically a special type of a statistical model (linear regression). By focusing on such models, this book provides a powerful, flexible, and up-to-date approach to data analysis.

1.3 Statistical Concepts

Because the present book focuses on regression analysis, it's helpful to be familiar with basic statistical concepts and terms. Don't worry: each model explored in this book will be interpreted in detail, and we will revisit most concepts as we go along. These are some of the most important terms we will use throughout this book: p*-values, effect sizes, confidence intervals,* and *standard errors.* This section provides a brief (and applied) review of these concepts. Bear in mind that we will simply review the concepts for now. Later we will revisit them in R, so don't worry if you don't know how to actually find *p*-values or confidence intervals: we will get to that very soon.

Before we proceed, it's important to review two basic concepts, namely, *samples* and *populations.* Statistical inference is based on the assumption that we can estimate characteristics of populations by randomly[1] sampling from them. When we collect data from 20 second language learners of English whose first language is Spanish (i.e., our sample), we wish to infer something about *all* learners who fit that description—that is, the entire population, to which we will never have access. At the end of chapter 2, we will see how to simulate some data and verify how representative a sample is of a given population.

Throughout this book, much like in everyday research, we will analyze samples of data and will infer population parameters from sample parameters. You may be familiar with the different symbols we use to represent these two sets of parameters, but if you're not, here they are: we calculate the sample mean ($\bar{x}$) to infer the population mean (μ); we calculate the sample standard deviation (s) to infer the population standard deviation (σ). Finally, our sample size (n) is contrasted with the population size (N).

1.3.1 p-*Values*

The notion of a *p*-value is associated with the notion of a **null hypothesis (H_0)**, for example that there's no difference between the means of two groups. *p*-values are everywhere, and we all know about the magical number: 0.05. Simply put, *p*-values mean the *probability of finding data at least as extreme as the data we have—assuming that the null hypothesis is true.*[2] Simply put, they measure the extent to which a statistical result can be attributed to chance. For example, if we compare the test scores of two groups of learners and we find a low (=significant) *p*-value ($p = 0.04$), we reject the null hypothesis that the mean scores between the groups are the same. We then conclude that these two groups indeed come from different populations, that is, their scores are statistically different because the probability of observing the difference in question *if chance alone generated the data* is too low given our arbitrarily set threshold of 5%.

How many times have you heard or read that a *p*-value is the *probability that the null hypothesis is true*? This is perhaps one of the most common misinterpretations of *p*-values—*the* most common on the list of Greenland et al. (2016). This interpretation is incorrect because *p*-values assume that the null hypothesis is true: low *p*-values indicate that the data is not close to what the null hypothesis predicted they should be; high(er) *p*-values indicate that the data is more or less what we should expect, given what the null hypothesis predicts.

Another common misconception about *p*-values is that they tell us the probability that we are making an error by rejecting the null hypothesis in a particular test. Naturally, this is not the case. As pointed out by Greenland et al. (2016), if the null hypothesis *is* in fact true for the test in question, then the probability of being in error is, well, 100%—not just 5%. Suppose you ran 100 studies, each of which contained one statistical test. Let's assume that all your 100 tests generated $p < 0.05$ and that you therefore rejected the null hypothesis 100 times. Given that our significance level (referred to as α) is 0.05, you would incorrectly reject the null hypothesis five times (Type I error).

Take another example. Suppose we simulate two different populations of learners. Population $\mathcal{A}$ has mean = 10, and population $\mathcal{B}$ has mean = 9—to keep things simple, both populations have the same standard deviation (σ = 3) and the same size of 100 learners. Because we are creating these populations ourselves, we know for a fact that their means are indeed different, that is, $\mu_{\mathcal{A}} \neq \mu_{\mathcal{B}}$—something we never actually know with real data. Now imagine that we simulated 100 random values for each population, respecting their different means, and that we ran a *t*-test. Finally, let's repeat this process 100 times, which means we would generate 100 random values 100 times, run 100 *t*-tests, and have 100 *p*-values. On average (because these simulations are random), we would fail to reject the null hypothesis between 30 and 40 times (!). In other words, we would conclude that the means are not different when in fact we know they are (**Type II error**).

Overall, *p*-values lead us to categorically infer that we either have a result (success) or we don't (failure)—see Kruschke (2015, ch. 11) for an in-depth analysis of *p*-values. This binarity, of course, largely underestimates the complexity of real data. Furthermore, it often drives us to do anything we can to achieve a significant *p*-value (e.g., ***p*-hacking**)—see, for example, Nuzzo (2014). Crucially, *p*-values tell us nothing about effect sizes—contrary to what some might think. If two groups are statistically different but their difference is so minuscule that it is irrelevant in practice, should we care about said difference?

As mentioned earlier, *p*-values measure the extent to which a statistical result can be attributed to chance. But, as noted by numerous researchers, they tell us nothing about our own hypotheses. If we reject the null hypothesis, how can we know the odds of the alternative hypothesis being correct? After all, very few people are actually interested in the null hypothesis *per se*.[3] Ultimately, if our own hypotheses are not plausible, the odds of their being correct will be low even if we do reject the null hypothesis—there could be multiple hypotheses which are compatible with the data and which are more likely to be true than an implausible hypothesis. Nuzzo (2014, p. 151) provides a useful example: if our initial hypothesis is unlikely to begin with (say, 19-to-1 odds against it), even if we find $p = 0.01$, the probability that our hypothesis is true is merely 30%.

Problems involving *p*-values have been known for decades—see, for example, Campbell (1982, p. 698). For that reason, many journals these days will require more detailed statistical results: whereas in the past a *p*-value would be sufficient to make a point (i.e., old statistics), today we expect to see effect sizes and confidence intervals as well (i.e., new statistics). That's one of the reasons that we will focus our attention on effect sizes throughout this book.

1.3.2 Effect Sizes

Effect sizes tell us how large the effect of a given variable is on the response variable. There are different ways to measure such effects. If you are familiar with *t*-tests and ANOVAs, you may remember that Cohen's d and η^2 are two ways to calculate effect sizes. In the models discussed in this book, effect sizes will be given as coefficients, or $\hat{\beta}$ values. The larger a $\hat{\beta}$ value is, the larger the effect size.

Here's a simple example that shows how effect sizes should be the center of our attention most of the time. Suppose you are interested in recording your classes and making them available to your students. That will require a lot of time and work, but your hypothesis is that by doing so, you will positively affect your students' learning. You then decide to test it, measuring their learning progress on the basis of their grades. You divide students into two groups. The control group (let's call it *C*) will not have access to recorded classes, whereas the treatment group (*T*) will. Each group has 100 students (our sample size), who were randomly assigned to either *C* or *T* at the beginning of the term. Assuming that all important variables are controlled for, you then spend the entire term on your research project.

At the end of the term, you analyze the groups' grades and find that the mean grade for group *C* was 83.05 and the mean grade for group *T* was 85.11. Let's assume that both groups come from populations that have the same standard deviation ($s = 3$). You run a simple statistical test and find a significant result ($p < 0.0001$). Should you conclude that recording all classes was worth it? If you only consider *p*-values, the answer is certainly *yes*—indeed, given that we are simulating the groups here, we can be sure that they do come from different populations. But take into consideration that the difference between the two groups was *only* 2.06 points (over 100)—sure, this tiny difference could mean going from a B+ to an A depending on how generous you are with your grading policy, but let's ignore that here. Clearly, the effect size here should at least make you question whether all the hours invested in recording all of your classes were worth it.

1.3.3 Confidence Intervals

Confidence intervals are likely one of the most abstract and misinterpreted concepts in traditional statistics. Assume we want to compare two groups and estimate the difference between them. We could simply collect one sample of data from each group and calculate the difference in means between the samples. But we don't know for sure the *real* difference between the two populations, because we are only working with samples. Confidence intervals give us a *range* of plausible values for the real difference based on the data we observe in our samples—this is preferred and more generalizable relative to a single number representing the difference in the samples.

If you repeated an experiment comparing two groups several times, each time would give you a slightly different difference in means, as well as a different confidence interval. Ninety-five percent of all such intervals would contain the true parameter value of interest (i.e., the true difference between the two populations under examination). You can see that the notion of confidence intervals rests on the assumption that you will *repeat* an experiment—which is not what typically happens in real life, unfortunately (hence the importance of replication studies). When we normally just run a single experiment, we cannot tell whether the only confidence interval that we have is the lucky interval to include the true parameter value.

Let's go back to our example earlier where we considered whether recording classes could be useful to students. The difference between the two groups, $\mathcal{T}$ and $\mathcal{C}$, was 2.06 points: $\bar{x}_{\mathcal{C}} = 83.05$ and $\bar{x}_{\mathcal{T}} = 85.11$. This difference was the effect size in our samples (i.e., the quantified impact of recording our classes using the original unit of the variable in question). What if we could have access to the *true* population means? Let's pretend we do: assume that the true difference between $\mathcal{T}$ and $\mathcal{C}$ is 1.98: $\mu_{\mathcal{T}} = 82.97$ and $\mu_{\mathcal{T}} = 84.95$ (in reality, of course, we wouldn't know these means). So the true difference in means is $\mu_{\mathcal{T}} - \mu_{\mathcal{C}} = 1.98$, which is not too far from 2.06, our sample means difference. As mentioned in §1.3.2, a t-test comparing both groups gives us a p-value < 0.0001, which means we reject the null hypothesis that the groups come from the same population. This is correct, since we generated them from different population means. The 95% confidence interval for the difference in means between the two groups is [1.13, 3.00].

The confidence interval in question does not include zero, so our p-value is less than 0.05. If we repeat the earlier steps 99 times, we will end up with 100 confidence intervals. Ninety-five percent of such intervals will contain the true parameter value—the difference in means between $\mathcal{T}$ and $\mathcal{C}$ here. Look again at the interval earlier. Does it contain the true parameter value? The answer is *yes*: the true difference, as we saw earlier, is 1.98 points, and 1.98 is in the interval [1.13, 3.00].

Confidence intervals are important because they provide one more level of information that complements the effect size. Because we are always dealing with samples and not entire populations, clearly we can't be so sure that the answer to our questions are as accurate as a single number, our effect size. Confidence intervals add some uncertainty to our conclusions—this is more realistic. Once we have our 95% confidence interval, we can examine how wide it is: the wider the interval, the more uncertainty there is. Wider intervals can mean different things; perhaps we don't have enough data, or perhaps there's too much variation in our data.

1.3.4 Standard Errors

To calculate a confidence interval, we need to know the *standard error of the sample mean* (*SE*), which is computed by dividing the standard deviation of the sample (s) by the square root of the sample size (n): $SE = \frac{s}{\sqrt{n}}$. Once we know the *SE*, our confidence interval is defined as $CI = [\bar{x} - 1.96 \cdot SE, \bar{x} + 1.96 \cdot SE]$[4]—later in this book we will use a function in R that calculates confidence intervals for us using a better method. When you collect data from a sample of participants, the mean of that sample ($\bar{x}$) will deviate from the true mean of the population (μ)—to which we have no access. As a result, there's always some degree of uncertainty when we infer the population mean from the sample mean. To estimate that uncertainty, we calculate the standard error of the sample mean.

The standard error is essentially the standard deviation of the sampling distribution of the sample mean. Let's unpack that. Imagine you collect test scores from five learners of English—so your sample size (n) is 5. This is a tiny sample of the entire population of *all* learners of English. You calculate the sample mean of the scores and you come to $\bar{x} = 84.2$. You then calculate the standard deviation of the sample (s), which in this hypothetical example is $s = 7.66$. As we know, the standard deviation quantifies the *variation* in the data. In our sample, students deviate 7.66 points from the mean in question (on average).

You now decide to repeat your data collection four times, where each time you collect scores from five different students. At the end, you will have five samples of the population of learners of English, each of which contains five scores. Each sample will in turn have its own mean and standard deviation. As a result, we will have five means. Assume that they are 84.2, 84.8, 77.4, 87.0, and 78.0. This is our *sampling distribution of the sample mean*. This distribution will be normal *even if* our population distribution is not normal, as long as the sample size is sufficiently large and the population has a mean (this is known as the Central Limit Theorem). If you compute the mean of these means, you will estimate the true mean of the population (μ). And if you compute the standard deviation of these means, you'll get the standard error of the sample mean, which quantifies the variation in the means from multiple samples. The larger the sample size of our samples (here $n = 5$), the lower the standard error will tend to be.[5] The more data you collect from a population, the more accurate your estimate will be of the true mean of that population, because the variation across sample means will decrease.

Given the example from the previous paragraph, you might think that the only way to estimate the standard error is to collect data *multiple* times. Fortunately, that is *not* the case. We saw in the first paragraph of this section that we can estimate the standard error even if we only have *one* sample by dividing the standard deviation of the sample by the square root of the sample size ($SE = \frac{s}{\sqrt{n}}$). This calculation works well assuming that we have a sufficiently large sample size. If that's not the case, we can alternatively *bootstrap* our standard error. Bootstrapping involves randomly (re)sampling from our own sample (instead of the population). This allows us to have a sampling distribution of the sample means even if our sample size is not ideal. We then take the standard deviation of that distribution, as described earlier. Finally, note that we can also calculate the standard error of other statistics (e.g., the median) by using the same methods.

In summary, while the standard deviation tells us about the variation of *raw data*, the standard error (from the mean) tells us about the estimated variation of *means*. Both statistics are informative, so we could in principle have both of them in a figure.

1.3.5 Further Reading

If you feel anxious about math in general and think you need to review basic statistical concepts in a little more detail, there are numerous options online these days. You may want to start with brief video tutorials, and then decide whether it's necessary to consult textbooks to understand different concepts in more detail. I recommend the following YouTube channels: Statisticsfun (http://www.youtube.com/user/statisticsfun/) and StatQuest with Josh Starmer (https://www.youtube.com/joshstarmer/). Both channels offer a wide range of short and intuitive videos on basic statistics.

You are probably already familiar with different statistics textbooks (there are hundreds out there), so you may want to try Wheelan (2013), which provides a more user-friendly take on important statistical concepts. I will make more specific and advanced reading suggestions throughout this book, once you're more familiarized with R. Finally, a recent and detailed review of key statistical concepts discussed earlier can be found in Greenland et al. (2016) and in numerous references therein—Greenland et al. provide all you need for the present book.

STATISTICAL CONCEPTS: SUMMARY

- You will never have to manually calculate standard errors in R, much like *p*-values, effect sizes, and confidence intervals
- It is certainly useful to know *how* to do so, but we won't get into such details here—R will calculate everything for us
- The main point here is to understand that most of what we observe in quantitative research in second language research relies on the concepts just reviewed earlier
- We will revisit them as we go through the different statistical models in this book (Part III)

Notes

1. In reality, of course, sampling is often *not* random (e.g., many of our participants are students on campus or live nearby).
2. And assuming that the *p*-values were computed appropriately.
3. Note that we cannot prove that the null hypothesis is true in Frequentist statistics. As a result, the absence of a difference between two groups *doesn't* prove that there is no difference.
4. You may remember that in a normal (Gaussian) distribution, 95% of the area under the curve lies within 1.96 standard deviations from the mean.
5. But note that we divide s by $\sqrt{n}$, not by n. Therefore, while it is true that SE is inversely proportional to n, doubling your sample size will *not* halve your SE.

2
R BASICS

This is the longest chapter in the book, as it provides an introduction to R. Don't assume that you will need to remember all the details from this chapter. Likewise, don't assume that you have to read the entire chapter in one sitting—you shouldn't do that. Read this chapter *slowly*, and make sure you practice the code presented here as much as you can in RStudio. In later chapters, you will likely want to come back here to review some of the fundamentals. That's absolutely normal: treat this chapter as your offline reference on the R language.

2.1 Why R?

R (R Core Team 2020) is a computer language based on another computer language called S. It was created in New Zealand by Ross Ihaka and Robert Gentlemen in 1993 and is today one of the most (if not *the* most) powerful tools used for data analysis. I assume that you have never heard of or used R and that therefore R is not installed on your computer. I also assume that you have little or no experience with programming languages. In this chapter, we will discuss everything you need to know about R to understand the code used in this book. Additional readings will be suggested, but they are not required for you to understand what is covered in the chapters to come.

You may be wondering why the book does not employ IBM's SPSS, for example, which is perhaps the most popular statistical tool used in second language research. If we use Google Scholar citations as a proxy for popularity, we can clearly see that SPSS was incredibly popular up until 2010 (see report on http://r4stats.com/articles/popularity/). In the past decade, however, its popularity has seen a steep decline. Among its limitations are a subpar graphics system, slow performance across a wide range of tasks, and its inability to handle large datasets effectively.

There are several reasons that using R for data analysis is a smart decision. One reason is that R is open-source and has a substantial online community. Being open-source, different users can contribute *packages* to R, much like different Wikipedia users can contribute new articles to the online encyclopedia. A package is basically a collection of tools (e.g., **functions**) that we can use to accomplish specific goals. As of October 2020, R had over 15,000 packages, so chances are that if you need to do something specific in your analysis, there is a package for that—naturally, we only need a fraction of these packages. Having an active online community is also important, as users can easily and quickly find help in forum threads.

Another reason that R is advantageous is its power. First, because R is a language, we are not limited by a set of preestablished menu options or buttons. If we wish to accomplish a goal, however specific it may be, we can simply create our own functions. Typical apps such as SPSS have a more user-friendly *Graphical User Interface* (GUI), but that can certainly constrain what you can do with the app. Second, because R was designed specifically for data analysis, even the latest statistical techniques will be available in its ecosystem. As a result, no matter what type of model you need to run, R will likely have it in the form of a package.

R is also fast, and most people would agree that speed is important when using a computer to analyze data. It is accurate to say that R is faster than any tool commonly used in the field of SLA.[1] This difference in speed is easy to notice when we migrate from software such as SPSS, which has a GUI, to a computer language, which instead has a command line. In R, we will rarely use our mouse: almost everything we do will be done using the keyboard.

Last but not least, R makes it easy to reproduce all the steps in data analysis, an advantage that cannot be overstated as we move towards open science. Reproducibility is also pedagogically valuable—whenever you see a block of code in this book, you can run it in R, and you will be able to follow all the steps in the analysis *exactly*. This efficiency in reproducing analytical steps is possible because R, being a language, relies on lines of codes as opposed to clicks on buttons and menus. The possibility to have a detailed script that contains all the steps you took in your analysis also means that you can go back to your study a year later and be sure to understand what exactly your analysis was doing.

In some ways, using R is like using the manual mode on a professional camera as opposed to using your smartphone: you may have to learn a few things about photography first, but a manual mode puts *you* in charge, which in turn results in better photos.[2]

2.2 Fundamentals

2.2.1 *Installing R and RStudio*

The first thing we need to do is install R, the actual programming language. We will then install RStudio, which is a powerful and user-friendly editor that uses R. Throughout this book, we will use RStudio, and I will refer to "R" and "RStudio" interchangeably, since we will use R *through* RStudio.

Installing R

1. Go to https://cloud.r-project.org
2. Choose your operating system

   - If you use *Linux*: choose your distro and follow the instructions[3]
   - If you use *Mac OS*: download the latest release (a .pkg file)
   - If you use *Windows*: click on base and download R (an .exe file) For help, visit https://cran.r-project.org/bin/windows/base/rw-FAQ.html

What you just installed is a combination of a programming language and an editor (you can see the editor is now on your computer). However, we will not use R's native editor, since it's not as powerful and user-friendly as RStudio. To install RStudio, follow these steps.

Installing RStudio

1. Go to https://rstudio.com and click on "Download RStudio"
2. Choose the free version and click "Download"
3. Under "Installers", look for your operating system

You should now have both R and RStudio installed on your computer. If you are a Mac user, you may also want to install XQuartz (https://www.xquartz.org) —you don't need to do it now, but if you run into problems generating figures or using different graphics packages later on, installing XQuartz is the solution. Because we will use RStudio throughout the book, in the next section, we will explore its interface. Finally, RStudio can also be used online at http://rstudio.cloud for free (as of August 2020), which means you technically don't need to install anything. That being said, this book (and all its instructions) is based on the desktop version of RStudio, not the cloud version—you can install R and RStudio and then later use RStudio online as a secondary tool. For reference, the code in this book was last tested using R version 4.0.2 (2020-06-22)—"Taking Off Again" and RStudio Version 1.3.1073 (Mac OS). Therefore, these are the versions on which the coding in this book is based.

2.2.2 Interface

Once you have installed both R and RStudio, open RStudio and click on File ≻ New File ≻ R Script. Alternatively, press Ctrl + Shift + N (Windows) or Cmd + Shift + N (Mac)—keyboard shortcuts in RStudio are provided in Appendix B. You should now have a screen that looks like Fig. 2.1. Before we explore RStudio's interface, note that the interface is virtually the same for Mac, Linux, and Windows versions, so while all the examples given in this book are based on the Mac version of RStudio, they also apply to any Linux and Windows versions of RStudio. As a result, every time you see a keyboard shortcut containing Cmd, simply replace that with Ctrl if you are using a Linux or Windows version of RStudio.

What you see in Fig. 2.1 is that RStudio's interface revolves around different panes (labeled by dashed circles). Panes B, C, and D were visible when you first opened RStudio—note that their exact location may be slightly different on your RStudio and your operating system. Pane A appeared once you created a new R script (following the earlier steps). If you look carefully, you will note that a tab called Untitled1 is located at the top of pane A—immediately below the tab you see a group of buttons that include the floppy disk icon for saving documents. Much like your web browser, pane A supports multiple tabs, each of which can contain a file (typically an R script). Each script can contain lines of code, which in turn means that each script can contain an analysis, parts of an analysis, or multiple analyses. If you hit Cmd + Shift + N to create another R Script, another tab will be added to pane A. Next, let's examine each pane in detail.

Pane A is probably the most important pane in RStudio. This is the pane where we will write our analysis and our comments, that is, this is RStudio's script window. By the end of this book, we will have written and run several lines of code in pane A. For example, click on pane A and write `2 + 5`. This is your first line of code, that's why you see `1` on the left margin of pane A. Next, before you hit enter to go to the next line, run that line of code by pressing Cmd + Enter. You can also click on the Run button to the left of Source in Fig. 2.1. You should now see the result of your calculation in pane B: `[1] 7`.

Pane B is RStudio's console, that is, it is where all your results will be printed. This is where R will communicate with you. Whereas you will write your questions (in the form of code) in pane A, your answers will appear in pane B—when you ran line 1 earlier, you were asking a simple math question in pane A and received the calculated answer in pane B. Finally, note that you can run code directly in pane B. You could, for example, type `2 + 5` (or `2 + 5` without spaces) in pane B and hit Enter, which would produce the same output as before. You could certainly use pane B for quick calculations and simple tasks, but for an actual analysis with several lines of code and comments,

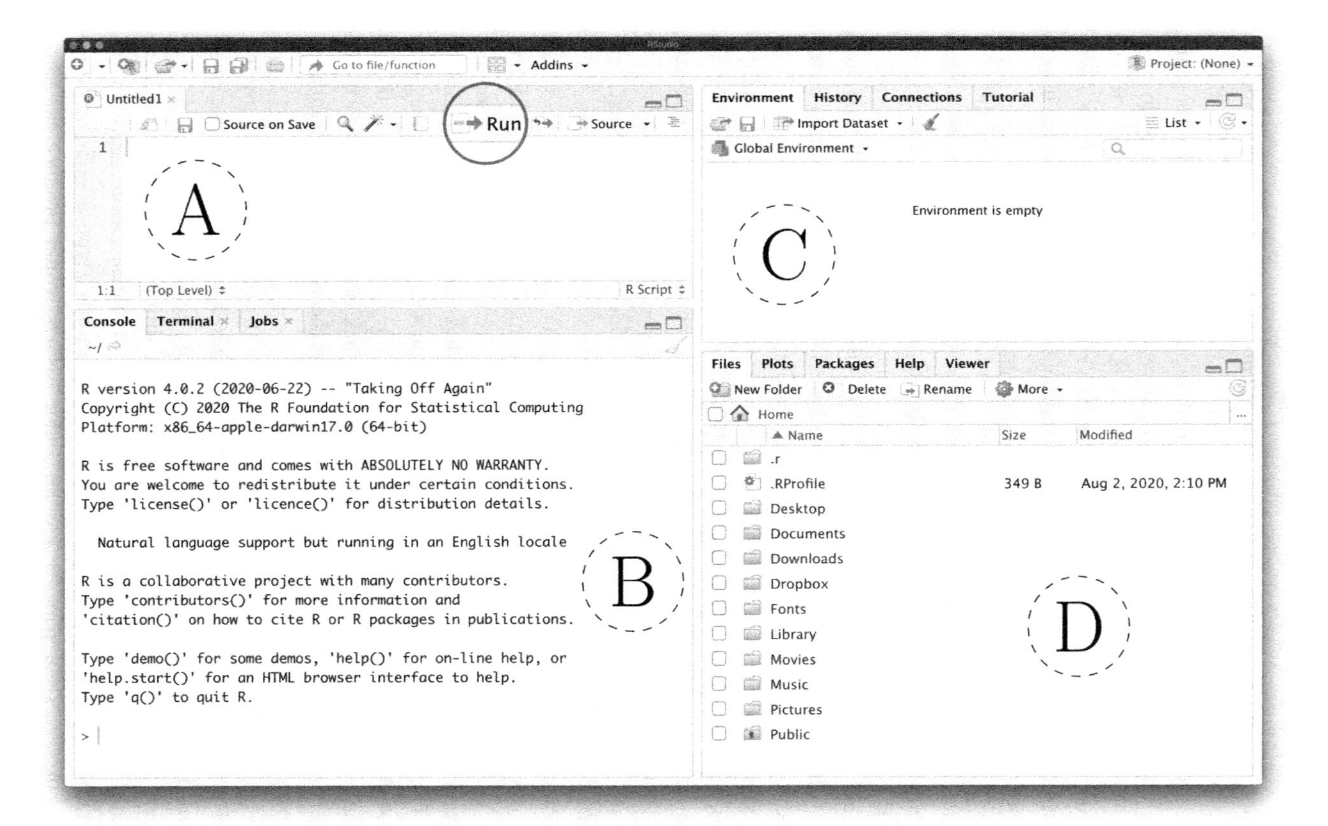

FIGURE 2.1 RStudio Interface and Pane Layout (Mac Version)

you certainly want the flexibility of pane A, which allows you to save your script much like you would save a Word document.

Pane C has three or four visible tabs as you first open RStudio—it can vary depending on the version of RStudio you are using, but we will not get into that in this book. The only tab we care about now is Environment. Pane C is where your variables will be listed (e.g., the objects we import and/or create in our analyses, such as our datasets). If that does not mean much to you right now, don't worry: it will be clear why that is important very soon.

Pane D is important for a number of useful tasks—there are five tabs in pane D in Fig. 2.1. The first tab (Files) lists the files and folders in the current directory. This is important for the following reason: R assumes that you are working in a particular folder on your computer. That folder is called **working directory**. If you wish to load some data, R will assume that the data is located in the current working directory. If it cannot find it there, it will throw an error on your console (pane B). You could either move your data to the working directory, or you could tell R exactly where your data file is located—we will see how to do that soon (§2.4).

The next tab in pane D is Plots. This is where all our figures will appear once we start working on data visualization in chapter 3. We then have Packages, which is a list of packages that we can use in our analysis. As mentioned before, R has thousands and thousands of packages, but most of what we will do in this book will just require a handful of powerful packages. We don't actually need to use this tab to install or load packages; we can do all that from panes A or B by writing some lines of code (§2.2.3).

Finally, pane D also has a tab called Help and a tab called Viewer. Whenever you ask for help in R by running `help(...)`, you will be taken to the Help tab. The Viewer tab is mostly used because RStudio allows us to work with several types of documents, not just R scripts. For example, you can use RStudio as your text editor and choose to produce a PDF file. That file would appear in the Viewer tab (or in a separate window)—we won't explore this pane in this book.

RStudio panes are extremely useful. First, they keep your work environment organized, since you do not need multiple windows floating on your screen. If you use R's native editor, panes A and B are actually separate windows, as is the window for your figures. RStudio, in contrast, organizes everything by adding panes to a single environment. A second advantage of RStudio's panes is that you can hide or maximize all panes. If you look again at Fig. 2.1, you will notice that panes A–D have two icons at the top right-hand corner. One icon looks like a box, and one looks like a compressed box. If you click on them, the pane will be maximized and minimized, respectively. For example, if you choose to maximize pane D, it will be stretched vertically, thus hiding pane C—you can achieve the same result by minimizing pane C. This flexibility can be useful if you wish to focus on your script (pane A) and hide pane B while you work on your own comments, or you could hide pane D because you are not planning to generate figures at the moment. These possibilities are especially important if you are working from a small screen.

If you like to customize your working environment to your liking, you can rearrange the panes in Fig. 2.1. Simply go to RStudio ≻ Preferences (or hit Cmd + ,) on a Mac. On Linux or Windows, go to Tools ≻ Global Options…. Next, click on Pane Layout. I do not recommend making any changes now, but later on you may want to make some adjustments to better suit your style. Finally, you can also change the appearance of RStudio by going to RStudio ≻ Preferences again and clicking on Appearance (on a Mac) or by going to Tools ≻ Global Options… ≻ Appearance (on Windows). Many people (myself included) prefer to write code on a darker background (one great example being the "Monokai" theme). If you choose Monokai from the list under Editor theme, you can see what that looks like. As with the panes, you may want to wait until you are comfortable with RStudio to decide whether you want to change its looks.

Now that you are familiar with RStudio's panes, we can really start looking into R. The next section will help you learn the fundamental aspects of the R language. We will explore everything you need to know about R to use this book efficiently.

2.2.3 R Basics

For this section, we will create one R script that contains a crash course in R, with code and comments that you will add yourself. First, let's create a folder for all the files we will create in this book. Call it bookFiles. Inside that folder, create another folder called basics—this is where we will work in this chapter. Second, make sure you create a new script (you can use the one we created earlier if you haven't closed it yet). In other words, your screen should look like Fig. 2.1. Third, save your empty script. Go to File ≻ Save (or hit Cmd + S on your keyboard). Choose an intuitive name for your file, such as rBasics, and save it inside basics, the folder you just created. RStudio will save it as rBasics.R—all R scripts are .R files. You will probably want to save your script every time you add a new line of code to it, just in case. Finally, all the code in question should be added to the newly created rBasics.R, so at the end of this section you will have a single file that you can go back to whenever you want to review the fundamentals.

The first thing we will do is write a comment at the top of our script (you can delete 2 + 5 from your script if it's still there). Adding comments to our scripts is important not only when we are learning R but also later, when we are comfortable with the language. In a couple of months, we will likely not remember what our lines of code are doing anymore. Even though some lines are self-explanatory, our code will become more complex as we explore later chapters in this book. Having informative comments in our scripts will help us understand what our past self did—and it will also help others reproduce our analyses. Comments in R must begin with a hashtag (#). Let's add a general comment at the top of our script (line 1), which will be a "title" for our script: # R Basics. If you try to run this line (line 1), R will simply print "# R Basics" in your console (pane B).[4]

> **FILE ORGANIZATION**
>
> There are ten code blocks in this chapter, which will be added to five different scripts—rBasics.R was our first script, so we will create four other scripts in this chapter. All five scripts should be located in your basics folder, which in turn is located in the bookFiles folder we created earlier. Remember to save your scripts after making changes to them (just like you would save any other file on your computer). Consult Appendix D if you wish to see how all the files in this book are organized or to which scripts you should add different code blocks.

Comments can also help with the overall organization of our R scripts. Some of the scripts we will create in this chapter (and in later chapters) will have multiple code blocks in them. For instance, code blocks 1, 2, and 3 should all go in rBasics.R. To keep the blocks visually separated inside the script, it's a good idea to add comments to the script to identify each code block for future reference. A suggestion is shown here, where a "divider" is manually created with a # and a series of equal signs (=====)—recall that everything that is to the right of a hashtag is a comment and so is not interpreted by R. We will discuss file organization again later (§2.4.2) and once more in chapter 3.

R code

```
1  # ========== CODE BLOCK X, CH. X
2
3  # code block content goes here
4
5  # ========== END OF CODE BLOCK
6
7  # ========== CODE BLOCK X, CH. X
8
9  # code block content goes here
10
11 # ========== END OF CODE BLOCK
```

SUGGESTION Keep Code Blocks Organized in Your Scripts

2.2.3.1 R as a Calculator

We saw earlier that R can be used as a powerful calculator. All the common mathematical operations are easy to remember: division (5 / 2), multiplication (5 * 2), addition (5 + 2), and subtraction (5 – 2). Other math operations include exponentiation (5 ** 2 or 5 ^ 2), modulus[5] (5 %% 2 = 1), and integer division (5 %/% 2 = 2). Let's add all these operations to our script and check their outputs. In Fig. 2.2, you can see our new script, rBasics.R, starting with a comment R basics in line 1. Line 2 is empty to create some white space, and line 3 creates a divider.

In line 5, we have a short comment explaining what the following line (6) does. This is obvious here, but it can save you some time in situations where the meaning of a function is not as apparent to you. Note that the cursor is currently in line 6. You can press Cmd + Enter anywhere in a given line and RStudio will run that line (it will also move to the next line automatically for you). Alternatively, you can also run multiple lines by selecting them and then pressing Cmd + Enter. In Fig. 2.2, I have already run all lines, as you can see in the console—note that the output of a comment is the comment itself. In the console, we will see, for example, that 5 %% 2 is 1 (the remainder of $5 \div 2$) and that 5 %/% 2 is 2 (the integer that results from $5 \div 2$). You don't have to reproduce the script in the figure exactly; this is just an example of how you could add some math operations and some comments to an R script.

Like with any decent calculator, you can also run more complex calculations. For example, choose any available line in rBasics.R and type 4 * sqrt(9) / (2 ** 5) – pi and run it (your output should be −2.766593)—as you open brackets, notice that RStudio will automatically close them for you. You can also type it in your console and hit Enter to run the code. Here we see two *functions* for the first time, namely, sqrt(9) ($\sqrt{9}$) and pi (π).

2.2.3.2 Variables and Vectors

The values generated by the calculations we have run so far are not actually stored anywhere. If you wanted to access the value of 5 * 2, the only way to do that would be to go back to line 9 and rerun it (or maybe scroll up and down your console window until you find the output you are looking for). This is not ideal—we need to learn about **variables** to be able to store and reutilize values.

Variables are objects that hold values in them. You can assign a piece of information to a variable and then access that piece of information by typing and running (i.e., "calling") the variable. In a way, we just did that—it is easier to remember to type pi than to remember a set of digits from π. Here are two ways of assigning values to a variable: x = 5 * 2 or x <- 5 * 2. Let us focus on using =, so type x = 5 * 2 in your script and run the line.

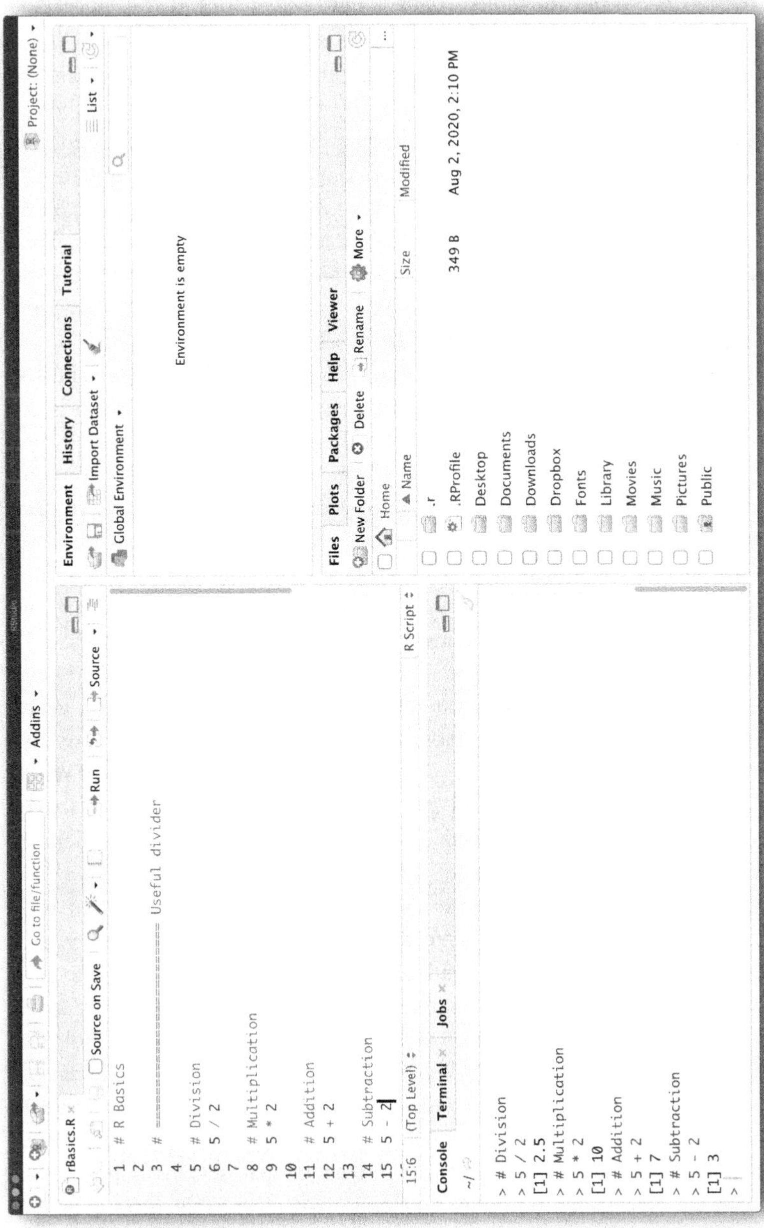

FIGURE 2.2 Starting Your First Script with RStudio: Math Operations

When you create a variable, you will notice that nothing is actually printed in your console (except the variable assignment itself). But if you now look at pane C (Environment), you will notice x is there, holding the value 10 (i.e., the result of 5 * 2). From now on, you can simply type x and run it, and R will print the value it holds (10). If you now type x = x + 10 in the next line of your script, you are telling R to "update" the value of x and which will now be 20 (x = 10 + 10). If you want to go back to what x was before, you can rerun the line where you specify x = 5 * 2 and *voilà*. By having different lines of code, you can go back to any stage of your analysis at any time. If you are used to Cmd + Z in other applications, having a script gives you all the redos you will ever need, with the advantage that you can choose which part of your analysis you wish to jump to by rerunning specific lines of code—the History tab in pane C will list all the lines of code we have already run.[6] This is illustrated in lines 1 and 2 in code block 1.[7]

Assigning simple operations or single numbers to variables is useful, but we need to be able to assign *multiple* values to a variable too. The problem is that if you run a line such as y = 1, 2, 3, you will get an error. To assign all three numbers to y, we need to use the c() function, which *concatenates* or combines different elements. If you run y = c(1, 2, 3) (line 4 in code block 1), you will now see that a new variable (y) has been added to your environment (pane C). As a result, if you simply type y in your script and run the line, R will print 1 2 3. As always, you can also type y in the console and hit Enter. Our new object, y, is a **vector**, which we've just created with the c() function.

Vectors are extremely important in R, as they allow us to hold however many values we need in a single variable. What if we want to have a vector that contains different words as opposed to numbers? Let's create a new variable, myWords, and assign three values to it: English, Spanish, French. Because these are words, we need *quotation marks* around each one, so you need to type myWords = c("English", "Spanish", "French")—this is demonstrated in line 6 in code block 1.

myWords is a vector that contains **strings**, or *characters*, as opposed to y, which contains only numbers. myWords also has a better name than y, as it conveys what the contents of the variable are—you should always try to have intuitive names for variables (x, for example, is not intuitive at all). Variable names cannot start with numbers, have spaces, or have special symbols that may already have a meaning in R—a hashtag, for example. But you can still use underscores, periods, hyphens and, crucially, lower- and uppercase letters (variable names are case-sensitive). myWords uses *camelcase*, where beginnings of non-initial words have capital letters instead of spaces. This is generally a good way to have short, intuitive, and easy-to-read variable names. Finally, you could rename y by reassigning it to a new variable: myNumbers = y. Now you have both variables in your environment, but you can forget about y and only use myNumbers instead (line 11 in code block 1).

Vectors have an important requirement: all the elements they contain must belong to the same class. As a result, you can have a vector with just numbers (like x) or a vector with just strings (myWords), but you cannot have a vector that contains numbers *and* strings at the same time. If you try doing that (mixedVector in line 8 in code block 1), R will force your numbers to be strings by adding quotes around the numbers ("1", "2", "English"). It does that because it cannot force strings to be numbers.

Now that we know what vectors are, we need to understand how we can access their contents. Naturally, you could simply type myWords and run it, and RStudio would print "English" "Spanish" "French" in your console. But what if you wanted to access just the first element in the vector? You can easily do that using what we call **slice notation**. To access the first element in myWords, simply type myWords[1]. The number within the square brackets represents the *index* of the element you want to access. What if you wanted both the first and the third elements in myWords? You can treat these two positions (indices) as numbers inside a vector and use that vector within the square brackets: myWords[c(1, 3)]. If you wanted the second and third elements, you could type myWords[c(2, 3)].

Besides accessing elements in a vector, we can also use different functions to ask a wide range of questions about the contents of said vector—lines 18–25 in code block 1. For example, length(myWords) will return 3, which is the number of items the vector contains. str(myWords) will return the class of the vector—I return to this function in §2.3. Because myWords contains strings, it is a *character vector*, abbreviated as chr in R if you run line 19 in code block 1. If you were to run str(myNumbers), you would get num, because all the members in myNumbers belong to the class numeric. Since we're dealing with a single vector, the same result can be achieved by running class (). Finally, another useful function is summary(), which will return different outputs depending on the object you wish to analyze. For the vectors in question, it will tell us about the length of the vector as well as its class. Later, when we explore more complex objects, summary() will provide more information.

In code block 1, we also see the function rev(), which reverses a vector. If myNumbers contains 1 2 3, then rev(myNumbers) will print 3 2 1. This also works with myWords, so running line 22 in code block 1 will give us French Spanish English. In line 23 in code block 1, we see rep(myWords, times = 2). Can you guess what this function is doing? It repeats the vector *n* times (here, *n* = 2). The output looks exactly like this: [1] "English" "Spanish" "French" "English" "Spanish" "French". Notice that all our outputs begin with [1]. That simply tells us that the first word that we see in that row is item number 1 in the vector. In other words, this is just R helping us visually count. While this may seem unnecessary for the vectors we are examining here, it can certainly be helpful when we deal with longer vectors, which involve multiple lines.

R code

```
x = 5 * 2          # x is now 10
x = x + 10         # x is now 20; run line 1 and it will be 10 again

y = c(1, 2, 3)     # a vector that concatenates three numbers

myWords = c("English", "Spanish", "French")

mixedVector = c(1, 2, "English")

# Renaming a variable:
myNumbers = y

# Slice notation:
myWords[1]         # = "English"
myWords[c(1, 3)]   # = "English" "French"

# Some useful functions:
length(myWords)
str(myWords)
class(myWords)
summary(myWords)
rev(myWords)
rep(myWords, times = 2)
mean(myNumbers)
sd(myNumbers)
```

CODE BLOCK 1 Variables and Vectors in R

The last two functions in code block 1 are mean() and sd(), for calculating the mean and the standard deviation of a numeric vector. Of course, these functions cannot be computed for character vectors such as myWords.

We have just explored vectors that contain two classes of objects, namely, character (strings) and numeric[8] (numbers). But these are not the only classes of objects we will encounter. For example, we have the logical class. As an example, let's say you want to find out whether 5 is greater than 2. You could type 5 > 2 and run it in R, and the answer (output) would be TRUE (note that *all* letters are capitalized). Likewise, if you type 5 == 5, the answer is TRUE: 5 equals 5. If you type 5 %in% c(1,3,5), the answer is also TRUE—you can read the %in% operator as "is present in".[9] Both TRUE and FALSE belong to the logical class, and so does NA ("Not Available", i.e., a missing value). Thus, we could add a vector to our collection of variables which contains elements from the logical class: myLogicals = c(TRUE, TRUE, FALSE, TRUE, NA). As we explore the chapters in this book, we will come across these (and other) classes, and you will see how they can be helpful in a typical analysis.

2.2.3.3 Lists

Thus far we have discussed vectors that contain numbers, strings, and logical values. As mentioned earlier, however, a vector cannot contain different classes of objects in it. Lists, on the other hand, can. As a result, we could have a list with *all* three classes we have examined earlier. To create a list, we use the list() function. Let's create a list with different classes of objects in it, and let's call it myList1—see line 2 in code block 2 (remember to add this code block to rBasics.R). myList1 contains three numbers (1, 2, 3), three strings ("English", "Spanish", "French"), and two logicals (TRUE, NA).

Notice that the numbers in the list are the same numbers in myNumbers and the strings in the list are the same strings in myWords. We could also create a list by typing list(myNumbers, myWords, c(TRUE, NA, FALSE))—see line 5 in code block 2, where this list is assigned to a new variable, myList2. These are seemingly identical lists, but if you run both of them and then call them separately, you will see that they are in fact structurally different. In myList1, items were manually and individually added to the list, whereas in myList2, all items are grouped together in separate vectors. As a result, if you run summary() on both lists (lines 7 and 8 in code block 2), you will see that myList1 has nine entries, each of which contains a single item (Length = 1). myList2, on the other hand, has only three entries, each of which contains three items (Length = 3). To actually see what the lists look like, run lines 10 and 11 in code block 2.

R code

```
1  # Each item is manually added to the list:
2  myList1 = list(1, 2, 3, "English", "Spanish", "French", TRUE, NA, FALSE)
3
4  # We use existing vectors to fill the list:
5  myList2 = list(myNumbers, myWords, c(TRUE, NA, FALSE))
6
7  summary(myList1)
8  summary(myList2)
9
10 myList1 # run this line to see what myList1 looks like
11 myList2 # then run this line: do you see the difference?
12
13 # Slice notation in lists:
14 myList2[[2]][3] # second entry, third item (= second item in myWords)
15
16 # Assign names to list entries:
17 names(myList2) = c("Numbers", "Languages", "Logicals")
18
19 # Now we can access all languages using our new names:
20 myList2[["Languages"]] # As opposed to myList2[[2]]
21 myList2$Languages      # The same can be achieved with "$"
```

CODE BLOCK 2 Lists in R

Lists look different from vectors. To understand them, think of a dictionary where we have different entries, and each entry can have different definitions. In myList2, the first entry is equivalent to myNumbers, which contains 1, 2, 3, and the second entry is equivalent to myWords, which contains "English", "Spanish", "French". Thus, if we wish to access "French", for example, we type myList2[[2]][3]—run line 14 in code block 2. This basically means "look inside myList2, take the second entry, and then give me the third element in that entry". When you run line 11 in code block 2, you will notice that each entry in a list starts with double square brackets [[ ]]. We can make things more complex by having a list whose first entry is a list whose first member is a list and so on.

Slice notation is useful, but we do not want to have to know which index to use. If we have an enormous list, we want to be able to access its content by using words, not numbers, as indices. We can do that by assigning names to a list's entries (see lines 17–21 in code block 2). When we have named the entries in our list, we can also access its contents by using "$"—line 21 in code block 2. We can also assign names to vectors the same way.

Lists are like dictionaries and can be very useful when we need a hierarchical structure that can hold objects that belong to different classes. When we explore statistical models in later chapters, we will use lists to explore their internal structure—you will notice that because some code blocks will have double square brackets in them. In addition, some objects in R are given in lists, so knowing how to work with lists can be important. But lists don't look like your typical spreadsheet, and your data will likely not be in lists. For that reason, we now move to data frames, the most important data structure that we use in the remainder of this book. You can always come back to this section on lists if you need to use lists, but ultimately you will use data frames much more often than lists.

2.3 Data Frames

Data frames resemble matrices (another data structure in R; run ?matrix()[10] to learn about it in your Help tab). But data frames have a very important characteristic that sets them apart—each column in a data frame can have its own class of objects. A data frame is roughly equivalent to an Excel spreadsheet. Unlike vectors, which have one dimension, data frames have *two* dimensions: rows and columns. Crucially, each column in a data frame is a vector.

Data frames are definitely not the fastest type of data structure; data tables, for example, are considerably faster (Dowle and Srinivasan 2019). However, they are the most popular, and unless you have a huge amount of data to analyze, data frames will be more than enough to get the job done.

You will probably not create a data frame in R. Instead, what typically happens is that you have some data in a file, for example, and you want to import it into R. That being said, data frames can also be useful when we want to create new data to explore the predictions of a statistical model—we will do this later on in Part III in this book. Let's take a look at a simple example that builds on the vectors we have already created. Here, we will make a data frame from myList2.

To create a data frame in R, we use the data.frame() command, as shown in code block 3[11]—remember to add this code block to rBasics.R. We then add column names and contents (every column in a data frame must have the same number of rows)—you can choose any name you want, but they must not start with special symbols, and they should not have spaces in them. Alternatively, because we want to have a data frame that has the exact content of myList2, we can use the as.data.frame() function (line 7 in code block 3)—but first we need to give our list entries names (line 6). We have already defined myNumbers and myWords in the same script (check to see whether these two objects/variables are in your Environment pane in RStudio. If you call (i.e., run) a variable, say, ABC, which no longer exists, you will get an error along the lines of Error: object 'ABC' not found. To avoid that, make sure you are still using the same script (rBasics.R) and that you have not closed RStudio in the meantime. If you have closed it, then rerun the lines of code where the variables are assigned and everything should work.

Our new variable, df, created in lines 1–3 *or* (6–)7 of code block 3, is a data frame that contains three columns. Each column has a different class of object, and each column has a name. We can now use a number of functions to better understand our data frame. For example, you may remember str() and summary() from code block 1. Both functions are very useful: str() will tell us the class of each column. If you run str(df), you will see all three columns preceded by $. We have already used dollar signs with lists to access different entries (as an alternative to using double square brackets). Data frames work the same way: always remember that if you want to access a column within a data frame, you first need to type the name of the data frame, then a dollar sign, and then the name of the column you want to access.

In df, column 1 is called Number and is a numeric column/variable (num); column 2, called Languages, is a character column (chr) with three different values in it; and column 3, called Logicals, contains values from the logical class (logi). Thus, if you wanted to print the Languages column on your screen, you would run df$Languages. Likewise, if you wanted to "extract" the column in question and assign it to a new variable, you would run newVariable = df$Languages—here we are simply copying (not removing) the column from df.

We can also use **slice notation** with data frames. For example, if you wanted to access the content of the Languages column, besides typing df$Languages, you could also type df[,2]. Because data frames have two dimensions (rows and columns), we have to specify both of them inside the

R code

```
df = data.frame(Numbers = myNumbers,
                Languages = myWords,
                Logicals = c(TRUE, NA, FALSE))

# Faster/simpler way in this case:
names(myList2) = c("Numbers", "Languages", "Logicals") # name list entries
df = as.data.frame(myList2)                            # create data frame from list

# Check class of each column:
str(df)

# Bird's-eye view of data frame:
summary(df)

# Calculate the mean, median, and standard deviation of Numbers column:
mean(df$Numbers)     # slice notation: mean(df[, "Numbers"]) or mean(df[,1])
median(df$Numbers)   # slice notation: median(df[, "Numbers"]) or median(df[,1])
sd(df$Numbers)       # slice notation: sd(df[, "Numbers"]) or sd(df[,1])

# Visualizing data frame:
head(df, n = 2) # top 2 rows
tail(df, n = 2) # bottom 2 rows

# Exporting a data frame as a csv file:
write.csv(df, file = "df.csv",
          row.names = FALSE,
          quote = FALSE)
```

CODE BLOCK 3 Creating a Data Frame in R

square brackets, in that order (df[row,column]). Therefore, df[,2] tells R to print all rows (no number before the comma) in column 2 (Languages). If we only wanted the third row in the second column, which would be equivalent to a single cell in an Excel spreadsheet, we'd type df[3,2]. If we wanted rows 1 and 3 in column 2, we would type df[c(1,3),2]—recall that if we want more than one item we must concatenate them in a vector. Finally, we could also use the *name* of a column instead of its index. Typing df[3,2] would therefore be equivalent to typing df[3, "Languages"]. Note that both df["Languages"] and df[, "Languages"] are interpreted by R, but they result in slightly different outputs. Whereas the latter prints only the *contents* of the column in question, the former prints the contents *and* the name of the column. Examples of slice notation for data frames can be found in code block 3 (lines 16–18).

Running summary(df) will give us a snapshot of the entire dataset—which is minuscule at this point. For example, if we have a numeric column, summary() will tell us the range, median, and mean of the column. You can also calculate the mean of any numeric column by using the mean() function. Lines 16–18 in code block 3 calculate the mean, median, and standard deviation of column Numbers in df—note that the dollar sign is used for all three functions.

One very important function that we have not discussed yet is head(). If we run head(df), R will print the first six rows in df (this is the default number of rows printed). Because df only has three rows, all three will be printed—line 21 explicitly adds the argument n = 2 to head(), so it will only print the top *two* rows. Another function, tail(), does exactly the same thing starting from the bottom of the data frame—see line 22 in code block 3. tail() will print the bottom rows of our data here (again, the default is n = 6).

Both head() and tail() are helpful because we will rarely want to print an entire data frame on our screen. To some people, this is likely one of the most striking interface differences between software like Excel and R. In R, we do not stare at our dataset at all times. Doing so is hardly ever informative, since some data frames will have too many rows and/or too many columns. Ideally, you want to know (i) what columns/variables you have as well as their classes (str()) and (ii) what your data frame looks like (head()). If you do want to have a spreadsheet-like view of your data frame, you can click on your data frame object in the Environment pane. You can also use the View() function (upper case V), which will open a new tab in pane A and show you the entire spreadsheet.

Finally, you may want to export the data frame we have just created as a csv file. To do that, use the write.csv() function. In lines 25–27 in code block 3, note that the function has four arguments[12]: first, we tell write.csv() what object (i.e., variable) we want to save. Next, we give the file a name (here, we want to save it as df.csv). Third, we use row.names = FALSE to tell R that we do not want an additional column with line numbers in our file. Lastly, we use quote = FALSE because we do not want R to use quotes around values (by default, quote = TRUE, which means every value in every row and column will be surrounded by quotes if you open the file in a text editor). The file will be saved in your current working directory—see the glossary. Later, we will see more concise options to export a data frame as a csv file.

Data frames are crucial, and chances are you will use them every single time you analyze your data in R. Thus, understanding how they work is key. In §2.5, we will explore another type of data structure, **tibbles**, which are very similar to data frames. Tibbles make our lives a bit easier, so they will be our focus throughout the book. Don't worry—everything we just discussed is transferrable to tibbles. Remember that most of the time you will *not* create a data frame within R. Rather, you will load your data into R, and the dataset will be interpreted as a data frame (or as a tibble, as we'll see in §2.5).

2.4 Reading Your Data

The very first thing we need to do to start a new analysis is to load our data. This, of course, assumes that your data is ready for analysis, which is not always the case—see §2.4.1. Most of the time, your dataset is a spreadsheet somewhere on your computer. Maybe you have an Excel file somewhere, and that is the file you wish to analyze. In this section, we will see how to load your file into R so you can start analyzing your data.

Before we actually import your data into R, however, we will discuss two important components of quantitative data analysis. First, we will check whether your data file is actually ready to be imported. Second, we will explore a powerful tool in RStudio that will help you keep all your projects organized.

2.4.1 Is Your Data File Ready?

In theory, an Excel file should contain *only* your data. Here's what that means: you have multiple columns, one observation per row, and all your columns have the same length (i.e., the same number of rows). The name of your columns should not contain spaces or special symbols. It is not a problem if you have empty cells, of course, but your file should not contain comments and notes besides the data, for example. Some people tend to write comments and notes in different cells in the same spreadsheet that they have their dataset. Other people also like to add formulæ to some cells, say, to calculate the mean of a given column. If that's your case, first copy *just* your data onto a new spreadsheet so that you have an Excel file that *only* contains your data and nothing else.

Once you have a file that only has data in it, you are ready to start. Even though R can certainly read Excel files (.xls), it is always a better idea to work with other file formats—.xls files will store not only your data but also charts and formulæ used in your spreadsheet, which are useless if we're using R for our data visualization and statistical analyses. In this book, we will use .csv files, which are plain text files where columns are separated by commas—hence the name *comma-separated values*. These files are lighter than .xls files and can be opened in any text editor. If your data is currently an Excel spreadsheet, simply save it as a .csv file.

DATA FILE

We will use a simple csv file to practice reading your data into R. Make sure you have downloaded sampleData.csv so you can follow every step.

2.4.2 R Projects

Whether you use SPSS or R, every research project that we develop has a number of files. Examples include folders for papers, reading materials, abstracts, and data files. Hopefully, all these folders are located in a single folder that gathers all the files that are related to a given research project. File organization is a good habit to cultivate, and RStudio offers us an incredibly handy tool for that: a file extension called .Rproj.

To understand what R Projects are, follow these steps. In RStudio, go to File ≻ New Project.... You will then have some options, two of which are New Directory and Existing Directory. As the names suggest, you should pick the former if you don't have a folder for a project yet and the latter in case you already have a folder where you want to place your data analysis files. We already created a directory earlier called basics, and that's where we will save our R Project. Therefore, choose Existing Directory and click on browse to locate the basics folder. Finally, click on Create Project. Your project will inherit the same name as the directory in which you create it, so it will be called basics.RProj. We will use this R Project for all the coding in the remainder of this chapter.

Once you have created your R Project, you will notice that RStudio will reappear on your screen. Only three panes will be visible (no script is open), so you can see your console, your environment, and pane D (from Fig. 2.1), where your Files tab is located. In that tab, you can see the contents of your newly created directory, where your R Project is located—you should be able to see only one file in the directory: basics.Rproj. You can confirm that this is the only file in the folder if you open that folder on your computer.

An Rproj file has no content in and of itself. It only exists to "anchor" your project to a given directory. Therefore, you could have multiple R Projects open at the same time, each of which would be self-contained in a separate RStudio session, so you would end up with multiple RStudios open on your computer. Each project would know exactly what directory to point to—that is another advantage of working with projects as opposed to single scripts. You do not necessarily need to use R Projects, but they can certainly help you manage all the files in your project. This book will use R Projects several times, and you're encouraged to do the same (your future self will thank you). I return to this point in chapter 3 (e.g., Fig. 3.1). Finally, you can place your rBasics.R file (created earlier for code blocks 1, 2, and 3) in the same directory as basics.Rproj, so there will be two files in the directory—you can delete df.csv, created in code block 3, since we won't use that file anymore.

2.4.3 Importing Your Data

We are now ready to import our data into R. The data in question (sampleData.csv) involves a very short hypothetical study. It consists of two groups of students (control and target) as well as three test scores for each student (testA, testB, testC)—there are ten students in the data, so our dataset has ten rows and five columns (10 by 5). This is a very common study design. For example, we could be examining the impact of two pedagogical approaches (target and control) on students' learning (as measured by test scores). We will only use sampleData.csv to practice importing files into R—in later chapters we will examine more realistic hypothetical data.

Place the file sampleData.csv (§2.4.1) in the directory where your .Rproj file is, which means your directory basics will now have three files (four if you count df, created in lines 23–25 in code block 3): one .R script (rBasics.R), one .csv, and one .Rproj. Next, start a new script by clicking on File ≻ New File ≻ R Script (the same steps from §2.2.2), or press Cmd + Shift + N to achieve the same result. Save your new script as dataImport.R, so that the file name is self-explanatory. You should now have four files in your directory.

There are several options to import sampleData.csv into R. One option is to use the function read.csv()—you may remember that we used write.csv() in code block 3 to export our data frame.[13] In your script (dataImport.R), write read.csv("sampleData.csv") and run the line to see what happens. You will notice that the entire dataset is printed in your console. But we want to assign our data to a variable, so that we can analyze it later. Let's name our variable ch2.

When you run ch2 = read.csv("sampleData.csv"), R will do two things: first, import the data file; second, assign it to a variable named ch2. As a result, even though the dataset is not printed in the console, a variable has been added to your environment. This is exactly what we want. Imagine reading a dataset with 1,000 rows and having the entire dataset printed in your console (!). Being able to see an entire dataset is only useful if the dataset is small enough (and that is almost never the case). Notice that ch2 is *not* a file—it's a variable inside RStudio. In other words, ch2 is a "virtual copy" of our data file; if we change it, it will not affect sampleData.csv. As a result, the actual data file will be safe unless we manually overwrite it by saving ch2 using write.csv (ch2, file = "sampleData.csv"), for example.

Once our variable ch2 has been created, we can use different functions to explore it. If you go back to code block 3, you will see a number of functions that we can now use to examine ch2. For example, we can run summary(ch2) to have a sense of what values each column contains as well as some basic statistics for numeric columns. We could also run str(ch2) to see the class of each variable (column). If you run it, you will notice that we have two columns that are chr: participant, which contains ten unique values (subject_1, subject_2, etc.), and group, which contains two unique values (control, target). Thus, we have ten participants and two groups in the data. We also have three num variables: testA, testB, and testC. Let's suppose that these are equivalent to a pre-test, a post-test, and a delayed post-test, for example.

Naturally, you could apply functions *directly* to specific columns. Recall that in a data frame, every column is a vector that can be accessed using a dollar sign

R code
```
# Import data file and assign it a variable: ch2
ch2 = read.csv("sampleData.csv")

# Summarizing the data:
summary(ch2)

# Checking variable classes:
str(ch2)

# Visualizing data frame:
head(ch2, n = 3)
# View(ch2) # opens new tab in pane A with dataset (remove first "#" to run this line)
```

CODE BLOCK 4 Importing Your Data into R

or slice notation. For example, if you wanted to calculate the mean of testA, you could run mean(ch2$testA). Finally, we can visualize the top and bottom rows of our data frame by using head() and tail(). By default, these functions will print six rows of data, but you can change that (see line 11 in code block 4).

2.5 The Tidyverse Package

Before we proceed, it's time to create another script. Even though you could do everything in a single script, it is useful to cultivate the habit of having one script for each type of task. For example, the script called dataImport.R has one main task: to import the data and check whether all is good.

Now that we have imported our data, let's create another script and save it as dataPrep.R. In this script, we will prepare the data for analysis. At the top of dataPrep.R, type and run source("dataImport.R"). When you run that line of code, R will run dataImport.R, and all the variables that are created within the script will appear in your environment (pane C). You can test it: click on the broom icon in pane C, which will remove all variables from your environment (you could also restart RStudio). Alternatively, you can type and run rm(list = ls()),[14] which will also remove all variables from your environment. Now run source("dataImport.R") and watch ch2 reappear in your environment.

You should now have a new script called dataPrep open. Next, let's install tidyverse (Wickham 2017), likely the most important R package you have to know about. tidyverse consists of a set of user-friendly packages for data analysis. Even though we could accomplish all our tasks without the packages in tidyverse, doing so would be more cumbersome and would require separate packages that do not necessarily have the same syntax. As we will see throughout this book, tidyverse makes R code more intuitive because of its more natural syntax, and you can do almost everything in this book using this collection of packages. Don't worry: by the end of the book, you will certainly be very familiar with tidyverse. Finally, you may recall that data tables were mentioned earlier (§2.3). If you'd like to use data tables instead of data frames (e.g., because you have too much data to process), you should definitely check the tidytable package (Fairbanks 2020). This package offers the speed of data tables with the convenience of tidyverse syntax, so you don't have to learn anything new.

To install tidyverse, we will use the function install.packages().[15] During the installation, you might have to press "y" in your console. Once the installation is done, we need to *load* the package using the function library(). The top of your script (dataPrep.R) should look like code block 5. Technically, these lines of code don't need to be at the top of the document; they must, however, be placed before any other lines that require them—overall, it is best to source, install, and load packages in the *preambles* of files. Finally, once a package is installed, you can delete the line that installs it (or add a hashtag to comment it out)[16]—this will avoid rerunning the line and reinstalling the package by accident. We are now ready to use tidyverse.

When you install and load tidyverse, you will notice that this package is actually a group of packages. One of the packages inside tidyverse is dplyr (Wickham et al. 2020), which is used to manipulate data; another is called tidyr (Wickham and Henry 2019), which helps us create organized data; another package is called ggplot2, which is used to create figures. We will explore these packages later—you don't need to load them individually if you load tidyverse.

2.5.1 *Wide-to-Long Transformation*

By now, we have created an R Project, an R script that imports sampleData.csv (which we called dataImport.R), and another script that prepares the data for analysis (dataPrep.R)—later we will import and prepare our data in a single script. When we source dataImport.R, we re-import our data variable, ch2. With that variable, we have used functions like summary(), str(), and head() to better understand what the structure and the contents of our data frame is. Our next step is to make our data *tidy*.

R code
```
1 # Script preamble: where you source scripts and load packages
2 source("dataImport.R")          # Runs everything in dataImport.R
3 install.packages("tidyverse")   # Comment this out once package is installed
4 library(tidyverse)              # Loads package
```

CODE BLOCK 5 Sourcing Scripts and Installing and Loading Packages in R

Throughout this book, we will rely on the concept of **tidy data** (Wickham et al. 2014). Simply put, a tidy dataset is a table where every variable forms a column and each observation forms a row. Visualize ch2 again by running head(ch2)—shown in Table 2.1. Note that we have *three* columns with test scores, which means our data is *not* tidy. This is not ideal because if we wanted to create a figure with "Test" on the *x*-axis and "Score" on the *y*-axis, we would run into problems. A typical axis contains information from *one* variable, that is, *one* column, but "Test" depends on three separate columns at the moment. We need to convert our table from a *wide* format to a *long* format. Wide-to-long transformations are very common, especially because many survey tools (e.g., Google Forms) will produce outputs in a wide format.

The data frame we want has a column called test and another column called score—shown in Table 2.2. The test column will hold three possible values, testA, testB, and testC; the score column will be a numeric variable that holds all the scores from all three tests. Let's do that using tidyverse, more specifically, a function called pivot_longer(). The discussion that follows will include code block 6, which you should place in dataPrep.R—see Table D.1 in Appendix D. You don't need to skip ahead to the code block yet; we will get there shortly.

TABLE 2.1 A Table in Wide Format

	participant	*group*	*testA*	*testB*	*testC*
1	subject_1	control	4.40	6.90	6.30
2	subject_2	control	6.50	9.90	6.10
3	subject_3	control	5.10	6.70	5.70
4	subject_4	control	4.60	9.60	5.50
5	subject_5	control	4.30	6.10	6.40
6	subject_6	target	6.90	8.80	5.10
...	...	...	...	...	...

TABLE 2.2 A Table in Long Format

	participant	*group*	*test*	*score*
1	subject_1	control	testA	4.40
2	subject_1	control	testB	6.90
3	subject_1	control	testC	6.30
4	subject_2	control	testA	6.50
5	subject_2	control	testB	9.90
6	subject_2	control	testC	6.10
...	...	...	...	...

Different functions have different requirements, also referred to as *arguments*. To use pivot_longer(), we first tell the function which columns we want to collapse. In ch2, these are testA, testB, and testC: c(testA, testB, testC). The second argument in the function, names_to =, asks for the name of the column that will hold the tests: names_to = "test". Finally, we choose the name for the column that will hold the scores (values_to = "score"). Alternatively, you can use the gather() function: gather(key = test, value = score, testA:testC).[17] This is all shown in code block 6.

Let's take a closer look at code block 6. You will notice a combination of three symbols, %>%, known as **pipe**—a percent symbol, a greater-than symbol, and a percent symbol again (you can type all three together by pressing Cmd + Shift + M). This operator is loaded by the magrittr package (Bache and Wickham 2014), which is in turn loaded when we load tidyverse. You can read this symbol as "and then". This is how we read the code in lines 2–5 in code block 6: "take variable ch2 *and then* apply the function pivot_longer()". By doing that, R knows that testA, testB, and testC are columns inside ch2. Finally, note that we are assigning the result of the function to a new variable, long. As a result, we still have our original data frame, ch2. We can also long-to-wide transform our data by using the pivot_wider() or spread() functions—see code block 6.

A nice thing about using %>% is that we can have a sequence of steps all at once. Technically, we don't have to use this symbol, but it makes our code much easier to read. Let's see a quick example: imagine that you wanted to create a sequence of numbers from 1 to 10, then take the average of the sequence, and then round up the number. Here's how you would normally do it: round(mean(seq(from = 1, to = 10))). Note that the first step, that is, creating the sequence, is the last function to appear from left to right. Conversely, the last step, rounding up the number, is the first function to be typed. In other words, the order in which you would describe what you are doing is the exact opposite of the order in which you would type your code. This can be confusing once we have multiple functions together. Using %>%, these steps are much easier to understand as they actually go from left to right: seq (from = 1, to = 10) %>% mean() %>% round(). Lastly, we don't need to add arguments to mean() or round() because they simply assume that the output of the previous step is the input they should use. Code block 6 summarizes all the functions discussed here (recall that you should add code block 6 to dataPrep.R, since long-to-wide and wide-to-long transformations are preparing your data for analysis).

Code block 7 runs head() on both the wide and long versions of our data—you can also see the R output that will be printed in your console, pane B. While ch2 is a 10-by-5 data frame, long is 30-by-4: long is therefore a *longer* data frame (30 rows and 4 columns).

R code
```
# From wide to long using pivot_longer():
long = ch2 %>%
  pivot_longer(c(testA, testB, testC),
               names_to = "test",
               values_to = "score")

# From wide to long using gather():
long = ch2 %>%
  gather(key = test,
         value = score,
         testA:testC)

head(long, n = 3) # Check result to see "test" and "score" columns

# From long to wide using pivot_wider():
wide = long %>%
  pivot_wider(names_from = test,
              values_from = score)

# From long to wide using spread():
wide = long %>%
  spread(test, score)

head(wide, n = 3) # Equivalent to ch2 (original data)
```

CODE BLOCK 6 Wide-to-Long and Long-to-Wide Transformations Using tidyverse

In summary, you can go from wide to long using either pivot_longer() or gather(), and you can go from long to wide using either pivot_wider() or spread()—all four functions belong to the tidyr package (loaded with tidyverse). Now that we have transformed our data frame from wide to long, we have one variable per column and one observation per row—we will discuss this type of data transformation once more in chapter 3 (§3.2).

We can finally start analyzing our dataset. We will use long as our main variable from now on.

2.5.2 Grouping, Filtering, Changing, and Summarizing Data

Once we have imported our data into R and prepared it for analysis, we will want to check certain patterns of interest. For example, we may want to check the mean score per group or perhaps remove certain data points or create a new column. tidyverse loads a package (dplyr) that allows us to accomplish all these tasks using intuitive functions. The main functions that we will use in this book are listed in Table 2.3.

R code

```
head(ch2, n = 3)  # BEFORE---output printed below:
#    participant       group testA testB testC
# 1    subject_1     control   4.4   4.9   8.4
# 2    subject_2     control   9.7   9.4   4.4
# 3    subject_3     control   5.1   7.9   7.9

head(long, n = 3) # AFTER---output printed below:
#   participant   group  test score
# 1   subject_1 control testA   4.4
# 2   subject_2 control testA   6.5
# 3   subject_3 control testA   5.1
```

CODE BLOCK 7 Data Frame before and after Wide-to-Long Transformation Using tidyverse

As discussed earlier, %>% allows us to apply multiple functions in a row. For example, if we wanted to know the mean test score (ignoring the three different tests) by group and then arrange our data from highest to lowest mean, we could use group_by(), summarize(), and arrange(), in that order and separated by %>%. Naturally, the very first step would be to point everything to our data, long—see code block 8.

Because each participant in our hypothetical study has three test scores, you may want to know the average test score by participant. To accomplish that, we use participant instead of group inside group_by(). Likewise, if you wanted to know the mean score by test, you would use group_by(test). Crucially, however, group_by() allows us to group our data by *multiple* variables. You could, for example, type group_by(participant, group), which would give you an output with three columns, namely, participant, group, and meanScore (which we create in the code shown in code block 8), which we can add to our dataPrep.R script.[18]

You may have noticed that the outputs of code block 8 is a tibble, not a data frame *per se*. Tibbles have a visual advantage over data frames: if you print tibbles on your screen, they will only show the first ten rows and only a number of columns that fit your screen. You can also use glimpse() to

TABLE 2.3 Main Functions in dplyr

Function	*Description*
arrange()	Orders data by a given variable
filter()	Filters data (creates a subset, i.e., removes rows)
group_by()	Groups data based on a specific variable
mutate()	Creates new column
select()	Selects columns in data (i.e., includes/removes columns)
summarize()	Summarizes data (typically used with group_by())

R code
```
# Mean scores by group:
long %>%
  group_by(group) %>%
  summarize(meanScore = mean(score)) %>%
  arrange(desc(meanScore))

# Mean scores by participant:
long %>%
  group_by(participant) %>%
  summarize(meanScore = mean(score)) %>%
  arrange(desc(meanScore))

# Removing (filtering out) controls:
targets = long %>%
  filter(group == "target") %>%
  droplevels()

# Removing column "group":
targets = long %>%
  filter(group == "target") %>%
  select(-group)

# Long-to-wide + new column (using pivot_wider()):
testDiff = long %>%
  pivot_wider(names_from = test,
              values_from = score) %>%
  mutate(difference = testC - testA) # Final score minus initial score
```

CODE BLOCK 8 Grouping, Summarizing, and Sorting Data with tidyverse

transpose the dataset and print all columns as rows. Data frames, on the other hand, will print *everything*, which is not very useful (that's why we often use head()). There are other small differences between these two objects, but they do not matter for now—just remember that even if you start out with a data frame, the output of summarize() in tidyverse will be a tibble. Throughout the book we will use tibbles, but you can treat tibbles and data frames as synonyms.

Next, let's see how we can remove certain rows, that is, filter our data. In lines 14–16 of code block 8, we are choosing to keep only participants in the target group in our data. We do that with filter(group == "target"), which is equivalent to filter(group != "control"): == means "is the same as"; != means "is different from". The other function used in lines 14–16 is droplevels(). This is why we're using that: remember that group is a factor with two levels, control and target. Now that we have chosen to keep only target participants, we still have two levels, but one of them (control) has no data points. Because of that, we may want to *drop* empty levels—this is like asking

R to forget that there ever were two groups in the data. That's what droplevels() does—and that's why it must come *after* we filter the data (order matters).

Instead of dropping the levels of group, we could actually remove the column from our data. If every participant now is in the target group, then we do not care about group anymore (it's redundant information). That's what lines 19–21 are doing in code block 8 with select(-group) (the minus sign subtracts the column from the tibble). If we also wanted to remove the participant column (for some reason), we would type select(-c(participant, group)). We need to concatenate them with c() because we now have more than one item in there.

Finally, let's see how to perform a more complex set of tasks. In lines 24–27 of code block 8, we are first transforming the test and score columns to a wide format (long-to-wide transformation)—see code block 6. Then, in line 27, we are creating a new column (mutate()) called difference, which will hold the difference (i.e., progress) between testC and testA. Notice that line 27 only works because its input is the output of pivot_wider(), so the actual input here is a tibble with testA, testB, and testC as separate columns again (much like ch2)—run testDiff to see what it looks like.

Being able to perform multiple actions using %>% can be incredibly useful, as we will see throughout this book. Crucially, we now have a variable in our script (testDiff) that contains a column with the difference between the two tests. As a result, you could use that particular column to create figures later on.

2.6 Figures

Once we have imported our data file and prepared our data for analysis, we can start generating some plots. Plots can help us visualize patterns of interest in the data *before* any statistical analysis. Indeed, we can often decide what goes into a statistical analysis by carefully examining different plots.

Before we generate our first plot, let's first create another script and save it as eda.R (*exploratory data analysis*).[19] This will be our third script (fourth, if you count rBasics.R) in the R Project created earlier (dataImport.R, dataPrep.R, eda.R)—see Appendix D. At the top of the script, we will source dataPrep.R, our previous step from earlier. Remember, by using source(), we don't actually need to open other scripts—remember to use "" around the file name before running source(). In fact, when you start typing the name of the script you want to source within source(), RStudio will show you a list of files that match what you are typing.[20] The next time you open your R Project, you can go directly to eda.R and run the line that sources dataPrep.R. You no longer have to run lines to import and prepare your data since those are run automatically when you source the scripts that contain those lines. Finally, because you have loaded tidyverse in dataPrep.R, you don't need to load it again in your eda.R.

In this section, we will generate a simple plot using one of the packages inside tidyverse called ggplot2. There are *several* packages that focus on data visualization in R, and you may later decide that you prefer another package over ggplot2—an incomplete list of relevant packages can be found at http://cloud.r-project.org/web/views/Graphics.html. However, ggplot2 is likely the most comprehensive and powerful package for plots out there. R also has its own base plotting system (which doesn't require any additional packages). Because every package will have some learning curve, the key is to select one package for data visualization and learn all you need to know about it. In this book, we will use ggplot2.

2.6.1 Using Ggplot2

Recall that long is a dataset with ten participants, two groups (control and target), three tests, and test scores. A natural question to ask is whether the scores in both groups are different.[21] For that, we could create a bar plot with scores on the *y*-axis and the two groups on the *x*-axis. We want bars (which represent the mean for each group) as well as error bars (for standard errors)—see §1.3.4. An example is shown in Fig. 2.3.

You should look at any plot in R as a collection of layers that are "stitched" together with a "+" sign. Each subsequent layer is automatically indented by RStudio and can add more information to a figure. The very first thing we need to do when using ggplot2 is to tell the package what data you need to plot. You can do that with the function ggplot(). Inside the function, we will also tell ggplot2 what we want to have on our axes. Let's carefully go over the code that generates Fig. 2.3, shown in code block 9.

In line 1, we source our dataPrep.R script (which itself will source other scripts). A month from now, we would simply open our R Project, click on our eda.R script and, by running line 1 in code block 9, *all* the tasks discussed earlier would be performed in the background. R would import your data, load

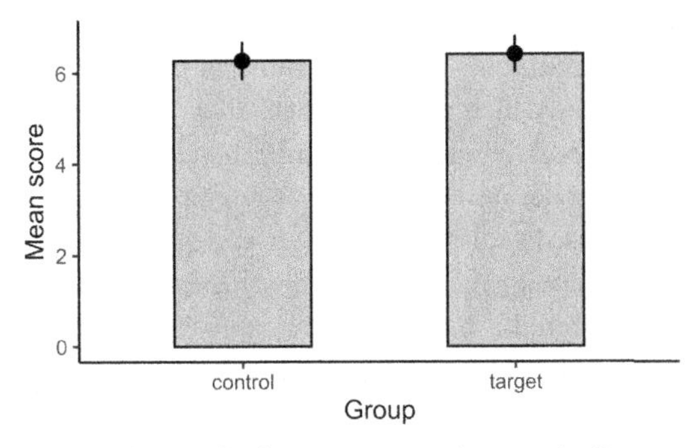

FIGURE 2.3 Bar Plot with Standard Error Bars Using ggplot2

R code

```
source("dataPrep.R") # This will load all the steps from earlier

# Your first ggplot:
ggplot(data = long, aes(x = group, y = score)) +
  stat_summary(geom = "bar",             # this will add the bars
               alpha = 0.3,
               color = "black",
               width = 0.5) +
  stat_summary() +                        # this will add the error bars
  labs(y = "Mean score",
       x = "Group") +
  theme_classic()
```

CODE BLOCK 9 Generating Your First Plot Using ggplot2

the necessary packages, and prepare the data, and we'd be ready to go. This automates the whole process of analyzing our data by splitting the task into separate scripts/components (which we created earlier). Chances are we won't even remember what the previous tasks are a month from now, but we can always reopen those scripts and check them out.

In line 4, we have our first layer, where we point ggplot2 to our data variable (long), and indicate what we want to have on our axes (aes(x = group, y = score)). The aes() argument is the *aesthetics* of our plot—it has nothing to do with the actual formatting (looks) of the plot, but with the *contents* of the axes. In line 5, we have our second layer—to improve readability, note that we can break lines after each +. We use a plus sign to tell ggplot2 that we're adding another layer to the plot (i.e., we are not done yet).

Line 5 uses a very important function: stat_summary(). This function is great because it provides a combination of data visualization and basic statistics. Inside stat_summary(), we have geom = "bar", which is simply telling ggplot2 that we want bars; alpha = 0.3, which is an optional argument, is adding some transparency to our bars (by default, bars are filled with solid dark gray). Transparency goes from alpha = 0 (transparent) to alpha = 1 (solid). The last two arguments of stat_summary() are color = "black" and width = 0.5, both of which are optional; they define the color of the borders of our bars and their widths. We then have another stat_summary() in line 9. This time, because we are not specifying that we want bars, ggplot2 will assume its default value, which is a point range (a dot for the mean and a line representing the standard error). Notice that we can add multiple layers of code to build a highly customizable figure.

Finally, the last two layers in our figure are labs() and theme_classic(). The former lets us adjust the labels in a figure, that is, rename the axes' labels in the figure. The latter is applying a theme to our figure; theme_classic() optimizes the figure for publication by removing the default light gray background, for example. Try running the code without that line to see what it looks like, but remember to also delete the + at the end of the previous line.

Note that we never actually told ggplot2 that we wanted to add standard errors to our bars. Instead, we simply typed stat_summary(). This is because ggplot2 (and R, more generally) will assume default values for certain arguments. One example is the function head(), which will display the first six rows of your data unless you explicitly give it another number (e.g., head (ch2, n = 10)). Like head(), stat_summary() assumes that you want standard errors from the means using a point range. If you want to something else, you have to explicitly indicate that within the function—that's what line 5 does in code block 9. When we run stat_summary(), what ggplot2 is actually interpreting is stat_summary(fun = mean_se, geom = "pointrange").

2.6.1.1 Commonly Used Plots

Besides bar plots, we often see scatter plots, box plots, histograms, and line plots. ggplot2 has functions that cover all such plots, and we will explore most of them throughout the book. ggplot2 often offers more than one function to achieve a given plot. To create our bar plot earlier, we used stat_summary(geom = "bar"), but we could have used geom_bar() as one of our layers instead (after your first layer, that is). To create a scatter plot, we will use geom_point(); to create a box plot, we will use geom_boxplot(). Don't worry about these functions now—we will go over them in detail as we explore the chapters in this book, especially the chapters in Part II.

2.6.1.2 Saving Your Plots

As with anything we do in R, there are different ways to save your plot. However, before saving it, we should create a folder for it in our current directory (basics)—let's call it figures.[22] One way to save plots created with ggplot2 is to use the function ggsave() right *after* you run the code that generates your plot. Inside ggsave(), we specify the file name (and extension) that we wish to use (file), and we can also specify the scale of the figure as well as the DPI (*dots per inch*) for our figure (dpi). Thus, if you wanted to save the plot generated in code block 9 to the figures folder, you'd add a line of code *after* line 12: ggsave (file = "figures/plot.jpg", scale = 0.7, dpi = "retina"). In this case, scale = 0.7 will generate a figure whose dimensions are 70% of what you can currently see in RStudio. Alternatively, you can manually specify the width and height of the figure by using the width and height arguments. To generate a plot with the exact same size as Fig. 2.3, use ggsave(file = "figures/plot.jpg", width = 4, height = 2.5, dpi = 1000).[23] If you realize the font size is too small in the figure, you can either change the dimensions in ggsave() (e.g., 3.5 × 2 instead of 4 × 2.5 will make the font look larger), or you can specify the text size within ggplot()—an option we will explore later in the book (in chapter 5). In later chapters, code blocks that generate plots will have a ggsave() line, so you can easily save the plot.

As mentioned earlier, you can run the ggsave() line after running the lines that generate the actual plot (you may have already noticed that by pressing Cmd + Enter, RStudio will take you to the next line of code, so you can press Cmd + Enter again). Alternatively, you can select *all* the lines that generate the plot plus the line containing ggsave() and run all of them together. Either way, you will now have a file named plot.jpg in the figures directory (folder) of your R Project.[24]

To learn more about ggsave(), run ?ggsave()—the Help tab will show you the documentation for the function in pane D. Formats such as pdf or png are also accepted by ggsave()—check device in the documentation. We will discuss figures at length throughout the book, starting in §2.6.2.

Finally, you can also save a plot by clicking on Export in the Plots tab in pane D and then choosing whether you prefer to save it as an image or as a PDF file. This is a user-friendly way of saving your plot, but there are two caveats. First, the default dimensions of your figure may depend on your screen, so different people using your script may end up with a slightly different figure. Second, because this method involves clicking around, it's not easily reproducible, since there are no lines of code in your script responsible for saving your figure.

2.6.2 General Guidelines for Data Visualization

There are at least two moments when visualizing data plays a key role in research. First, it helps us understand our own data. We typically need to *see* what's going on with our data to decide what the best analysis is. Naturally, we can also use contingency tables and proportions/percentages, but more often than not figures will be the most appropriate way to do that. For example, if we want to verify whether our response variable is normally distributed, the quickest way to do that is to generate a simple histogram.

The second moment when visualizing data is crucial is when we communicate our findings in papers or at conferences. The effectiveness of a great idea can be compromised if it's not communicated appropriately: if the reader (or the audience) cannot clearly see the patterns on which your analysis depends, your study may come across as less convincing or less meaningful. Furthermore, showing your data sends a message of transparency: if I can't *see* your data or the patterns to which you refer, I might wonder why you're not showing them to me.

These observations may seem obvious, but a great number of papers and presentations seem to ignore them. Overall, every study that uses data should show the patterns in the data. And because most studies in second language research rely on data, it wouldn't be an exaggeration to assume that nearly all studies in the field should take data visualization seriously.

Now that you've seen a brief demonstration of how a figure can be generated in R, let's focus on some key conceptual aspects involving data visualization. Later, in chapters 3–5, we will examine how to create figures in R in great detail.

2.6.2.1 Figures should be Consistent with the Statistical Analysis

Your analysis should be internally consistent, and so your figures should organically lead the reader to your statistical analysis. What this means is that by the time the reader reaches your statistical analysis, you have already presented the variables of interest (in figures, tables, or text).

Your figure should also present the data using the same perspective used in the statistical analysis. For example, your y-axis should be the response variable that you will later use in your statistical analysis, and your x-axis should be one of your explanatory variables. Imagine that you want to test how different teaching methods affect students' scores on a particular test. Their scores should be the y-axis in your figures, and it should also be the response variable in your statistical analysis. Not all variables deserve a figure, of course, and you should focus your data presentation on the variables that play an important role in your study. The important thing is to treat your figures and your statistical analysis as *consistent* and complementary parts of the same whole.

2.6.2.2 Figures should Communicate Information with Clarity

It's often the case that less is more. 3D figures, for example, can certainly look nice—especially if you can rotate them using your mouse. But on a two-dimensional sheet of paper, where everything is static, they can look very messy. Carefully consider the amount of information you present in a figure, and ask your colleagues if they think the message in the figure is clear. For example, to communicate patterns more effectively, you might sometimes have to visualize only a (representative) subset of your data—depending on the plot, using all your data points may result in a figure that contains too much information.

When it comes to creating an appropriate figure, there is clearly a sweet spot: you don't want to make a figure that simply shows a couple of percentage points. That will look silly and will waste a lot of space. At the same time, you don't want to overpopulate your figure with too many variables (or data points, as mentioned earlier). In many situations, you should aim for more than two levels of information—but it all depends on *what* those levels are and how you plan to explore them. We will discuss this at length throughout the book.

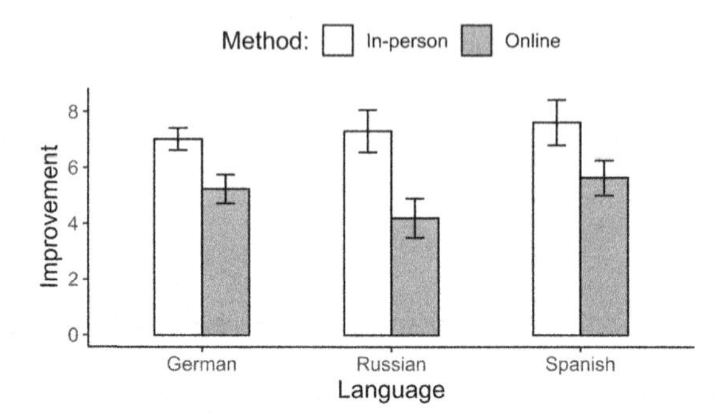

FIGURE 2.4 Three Variables in a Bar Plot Using ggplot2

Take the famous bar plot that only shows mean values and nothing else. Most of the time, you don't need a figure to communicate only means, naturally. If you plan to use bar plots, you should always add error bars (if they make sense) and ideally have at least one more dimension to visualize. In a study that explores different teaching methods and their impact on learning English as a second language (e.g., teaching in-person *vs.* online), you will likely have more than one native language in your pool of participants (by design). Your bar plot could then have along its *x*-axis the languages in question, and the fill of the bars could represent the teaching methods of interest—see Fig. 2.4.

2.6.2.3 Figures should be Aesthetically Appropriate

It's surprisingly common to see figures that are pixelated, whose labels are too small (or too large) relative to the text, or whose font family is inconsistent with the text. Another common issue (especially in presentations) is the choice of colors: if you have a dark background in your figure, the choice of colors must take that into account, otherwise it will be much harder for the audience to read your slides—especially considering that projectors often don't have great contrast and resolution.

Your figures should look crisp even when you zoom in, and they should not distract the reader from what really matters (again: less is more). You should have clear labels along the axes (when needed) and an informative caption, and your use of color should serve a purpose—remember that using colors can actually be a problem in some journals. In reality, it's usually possible to remove the colors from a figure and use some other dimension to convey the same information—we will exercise that throughout the book.

Fig. 2.4, like the figure we discussed earlier in this chapter (Fig. 2.3), shows a bar plot. Here, however, we have three dimensions: the *y*-axis shows the change in score (e.g., pre- and post-test) of the hypothetical study mentioned earlier; the *x*-axis shows the different native languages of the participants in said study; and the fill of the bars discriminates the two methods under examination. Note that the font sizes are appropriate (not too small, not too large). In addition, the key is positioned at the top of the figure, not at the side. This allows us to better use the space (i.e., increase the horizontal size of the figure itself). No colors are needed here, since the fill of the bars represents a factor with only *two* levels ("In-person" and "Online"). Finally, the bars represent the mean improvement, but the plot also shows standard errors from the mean. If the plot only showed means, it would be considerably less informative, as we would know nothing about how certain we are about the means in question (i.e., we would not incorporate information about the variance in the data).

The take-home message from Fig. 2.4 is that an effective figure doesn't need to be fancy or colorful. The reader, in this case, is not distracted by small font sizes or too many colors that don't represent any specific values in the data. Instead, bars are grouped together in an intuitive way: each language group contains one bar for in-person classes and one bar for online classes, which allows the reader to easily compare the effects of the methods in question for each language group. Imagine now that the *x*-axis represented instead the teaching methods and the fill color represented the different native languages. In that case, the reader would need to look *across* categories on the *x*-axis for each language to visually estimate the effect of the main factor (method).

Assuming what was discussed in §2.6.2.1, Fig. 2.4 implies that the statistical analysis will assess the effects of native language and method on learners' improvement. On top of that, the figure suggests that teaching method matters (in-person being more effective than online teaching in this hypothetical example) and that its effects are relatively similar across all three languages in question. All these pieces of information will help the reader understand what is likely going on *before* the actual statistical analysis is presented and discussed.

2.7 Basic Statistics in R

So far in this chapter we have covered *a lot*. We installed R and RStudio, discussed RStudio's interface, and explored some R basics with a number of code blocks, which led us to ggplot2 and figures in the previous section. Once we have a figure, the natural next step is to statistically analyze the patterns shown in said figure. Thus, in this section of the chapter, we will turn to some basic statistical concepts—this will be a good opportunity to review some concepts that will be used throughout the book. We'll first start with one of the most basic concepts, namely, *sampling*.

Assume we want to collect data on proficiency scores from 20 learners. To simulate an entire population of learners containing, say, 1 million data points, we can use the function rnorm(), which *randomly* generates numbers following a normal (Gaussian) distribution[25] (there's no reason to assume that proficiency scores are not normally distributed here). Create a variable called pop and assign to it rnorm(n = 1000000, mean = 85, sd = 8)—you can even do this in the console, without creating a new script, since it will be a simple exercise. This will generate one million scores normally distributed around a mean (μ) of 85 points with a standard deviation (σ) of 8 points. In reality, we never know what the mean and standard deviations are for our population, but here we do since we are the ones simulating the data. You can check that both μ and σ are roughly 85 and 8, respectively, by running mean(pop) and sd(pop)—you won't get exact numbers, but they should be very close to the parameter values we set using rnorm().

Next, let's sample 20 values from pop—this is equivalent to collecting data from 20 learners from a population of one million learners. To do that, create a new variable, sam, and assign to it sample(x = pop, size = 20). Now, run mean() and sd() on sam, and you should get a sample mean ($\bar{x}$) and a sample standard deviation (s) that should be very similar to the *true* population parameters (μ, σ) we defined earlier. This is a quick and easy way to see sampling at work: we define the population ourselves, so it's straightforward to check how representative our sample is.

We have now seen how to simulate some data using R, and we have also seen how to load our data, adjust it for analysis, and visualize the main pattern of interest using a bar plot. This is the typical order you will follow every time you have a project that involves data analysis. The final step in that process, running some stats on the data, is the topic of this section. You can probably guess what the first step is: create a new script and save it as stats.R. This is the script where we will review basic statistical concepts and learn how to use R to perform simple tests. You should now have four separate scripts in the folder where you have your R Project: dataImport.R, dataPrep.R, eda.R, and stats.R (plus rBasics.R and a folder for figures called figures, where you saved plot.jpg earlier). At the top of stats.R, you should type and run source("eda.R")—which means next time you open the basics.Rproj you can go directly to stats.R if that's the file you need (i.e., it will automatically load the previous scripts associated with the project). Let's begin by revisiting some key concepts used in traditional (Frequentist) statistics. We will do that by using our long dataset.

2.7.1 What's Your Research Question?

If you already have actual data, that means you have long passed the stages of formulating a *research question* and working on your *study design*. Having a relevant research question is as important as it is difficult—see Mackey and Gass (2016, ch. 1). Great studies are in part the result of great research questions. Your question will dictate how you will design your study, which in turn will dictate how you will analyze your data. Understanding all three components is essential. After all, you may have an interesting question but fail to have an adequate study design, or you may have both a relevant question and an appropriate study design but realize that you don't know how to analyze the data.

The hypothetical dataset in question (long) consists of two groups of participants (control and target) as well as three sets of test scores—we will examine more complex and realistic datasets later on, but for now our tibble long will be sufficient. Suppose that our study examines the effects of a particular pedagogical approach on students' scores, as briefly mentioned earlier. For illustrative purposes, let's assume that we want to compare a student-centered approach (e.g., workshop-based), our targets (treatment group), to a teacher-centered approach (e.g., lecture-based), our controls, so we are comparing two clearly defined approaches (two groups). Next, we must be able to measure scores. To do that, we have decided to administer three tests (e.g., a pre-test (testA), a post-test (testB), and a delayed post-test (testC), as alluded to earlier). With that in mind, we can ask specific questions: is the improvement from testA to testB statistically different between controls and targets? How about from testA to testC? These questions must speak directly to your overarching research question. Finally, in addition to a research question, you may have a directional hypothesis, for example, that a lecture-based approach will lead to *lower* scores overall, or a non-directional hypothesis, for example, that there will be some difference in learning between the two approaches, but which one will be better is unknown. Let's now go over some basic stats assuming the hypothetical context in question.

2.7.2 t-Tests and ANOVAs in R

Likely the most popular statistical test known to us all, a *t*-test compares two means ($\bar{x}$) and tests whether they are statistically different. If they are statistically different, then we reject the null hypothesis that both groups come from the same population. We could run a *t*-test comparing our two groups (controls and targets), or we could compare the mean of a single group with a baseline mean of interest (e.g., 75)—say, to verify whether the mean of group X is statistically different from 75. Either way, we are comparing two means.

If we wish to compare more than two means, then we typically perform an analysis of variance, or ANOVA. There are many types of ANOVAs (e.g., ANCOVA, MANOVA), but here we will simply explore a one-way and a two-way ANOVA, so you can see how to run ANOVAs using R.

Both *t*-tests and ANOVAs are used when you have a continuous and normally distributed (Gaussian) dependent variable (score in long). A *t*-test requires *one* independent[26] variable (e.g., group in long), while an ANOVA can have multiple variables. We will later refer to these as the *response* variable and the *predictor* or *explanatory* variable, respectively, to better align our terminology with the statistical techniques we will employ.

Our tibble, long, has a continuous response variable (score). However, our scores do *not* follow a normal distribution (an assumption in parametric tests). You can see that by plotting a histogram in R: ggplot(data = long, aes (x = score)) + geom_histogram()—we will go over histograms in more detail in chapter 3, more specifically in §3.3. Clearly, we *cannot* assume normality in this case.[27] Here, however, we will—for illustrative purposes only. R, much like any other statistical tool, will run our test and will not throw any errors: it is up to us to realize that the test is not appropriate given the data.

To run a *t*-test in R, we use the t.test() function—code block 10 (line 4). This will run a Welch two-sample *t*-test—which, by default, is *two-tailed* (i.e., is non-directional) and assumes unequal variances (i.e., that both groups don't have the same variance).[28] If we have a directional hypothesis, we can run a *one-tailed* t-*test* by adding alternative = "greater" if your assumption is that the difference in means is greater than zero. You could also use alternative = "less" if you assume the difference is *less* than zero. By default, *t*-tests assume alternative = "two.sided" (i.e., two-tailed).

To run an ANOVA, we use the aov() function (code block 10 (line 13)). Naturally, you can assign these functions to a variable, so the variable will "save" your output as an object in your environment (see line 16). Like any other function in R, both of them require certain arguments. In addition, you must remember a specific syntax to be able to run these functions: $Y \sim X$, where Y is our response variable and X is our explanatory variable. In code block 10, the *t*-test and ANOVA are run directly. If you examine line 13, you will notice that we are running the ANOVA inside the function summary(). This is because aov() will not give us the output that we typically want—run the line without summary() to see the difference.

The output of a typical *t*-test in R will include a *t*-value, degrees of freedom, and a *p*-value. It will also state the alternative hypothesis (i.e., that the true difference in means is not equal to zero) and the 95% confidence interval (CI). Finally, it will show us the means of each group being compared in the data in question. The output of a typical ANOVA, on the other hand, will not print a *p*-value—that's why summary() is used in code block 10, which provides not only a *p*-value but also an *F*-value. It will, however, print the sum of squares, the degrees of freedom, and the residual standard error. This should all be familiar, but don't worry too much if it isn't—we won't be running *t*-tests and ANOVAs in this book.

You may have already figured out that we won't be able to answer important questions with a simple *t*-test without making some adjustments to our data. First of all, if we run a *t*-test on the scores of the two groups of participants, we can't accommodate the different test scores into our test. In other words, we are treating the scores as if they came from the same type of test. This is exactly what Fig. 2.3 conveys: it ignores that we have three *different* tests and only focuses on the two groups of participants that we have in our study. Clearly, this is not ideal.

Imagine if we had a column in our data that subtracted the score of the pre-test from the delayed post-test. This new column would tell us how much improvement there was between these two tests—positive values would mean a student improved. The problem, however, is that to create such a column, we need pre-test and delayed post-test scores to be their own column, which means our long tibble will not work. We have two options: go back to ch2, which is a wide tibble, or long-to-wide transform long. Since ch2 should be loaded in your environment, we can add our new column to it using the mutate() function in lines 7–8 in code block 10. This technique simplifies our data to some extent: we go from two tests to the difference between them—note that we have not removed any columns from ch2. This, in turn, allows us to indirectly "compress" two tests of the variable test in a *t*-test that already has its explanatory variable defined (group) and therefore wouldn't have room for yet another variable (test). Our response variable now contains information about any potential improvement between testA and testC.

If we wanted to check whether the mean scores are different between tests without simplifying our data, we could run a one-way ANOVA, which would then ignore group as a variable. If we instead wanted to consider both group and test, we would run a two-way ANOVA—both examples are provided in code block 10.

2.7.3 A **Post-Hoc** *Test in R*

When we run an ANOVA such as twoWay in code block 10 (score ~ test + group), we typically have a variable with more than two unique values in it. Here, the variable test has three levels (i.e., it's equivalent to a factor with three levels if you transform its class from chr to Factor). As a result, if our *p*-value is significant (which it is), we must find out where the difference is. Line 19 in code block 10 is helping us address that question by running a specific *post-hoc* test (**Tukey HSD**)—its output is shown in lines 20–24. Let's go over lines 16–19 in detail.

The first thing we want to do if we plan to use our ANOVA output elsewhere (for a plot, for example), is to assign it to a variable, as usual. That's what line 16 is doing. Next, we can use the function summary() to inspect the results of our two-way ANOVA. You will notice that each explanatory

R code

```
source("eda.R")

# T-test: t(27.986) = -0.25, p > 0.1
t.test(score ~ group, data = long)

# T-test comparing score differences (C-A) between groups:
ch2 = ch2 %>%
  mutate(AtoC = testC-testA)

t.test(AtoC ~ group, data = ch2)

# One-way ANOVA: F(2,27) = 15.8, p < 0.001
summary(aov(score ~ test, data = long))

# Two-way ANOVA:
twoWay = aov(score ~ test + group, data = long)
summary(twoWay)

TukeyHSD(twoWay, which = "test")
# $test
#              diff        lwr        upr     p adj
# testB-testA  2.58  1.3159753  3.8440247 0.0000802
# testC-testA  0.32 -0.9440247  1.5840247 0.8055899
# testC-testB -2.26 -3.5240247 -0.9959753 0.0004168

plot(TukeyHSD(twoWay, which = "test"))
```

CODE BLOCK 10 Running *t*-Tests and ANOVAs in R (Results Given as Comments)

variable of interest has its own line. Line 19 runs our *post-hoc* test using TukeyHSD(). The which argument inside TukeyHSD() exists because we have *two* variables of interest here, but only one has more than two levels (test). In lines 22–24, we see all three pairwise comparisons (testB − testA, testC − testA, and testC − testB), their differences (diff), the lower (lwr) and upper (upr) bounds of their 95% CI, and their adjusted *p*-value (to reduce the probability of Type I error as a result of multiple comparisons).

Because the output of our *post-hoc* test has a lot of numbers in it, it may be better to visualize the comparisons of interest. This brings us to line 26, which plots the results using the plot() function—note that here we are not using ggplot2. Line 26 will generate a figure that contains the output shown in lines 20–24, but unlike line 19, it will not print an output. This is because plot(), being the first function from left to right, dictates what R will do.[29]

In the figure generated by line 26 in code block 10, all three comparisons are shown along the *y*-axis.[30] Two of the three comparisons are significant—those printed in lines 22 and 24 of code block 10. The *x*-axis shows the actual differences in mean as well as 95% CI. If a confidence interval for a given comparison includes zero, the *p*-value of said comparison will be greater than 0.05 and therefore not significant based on the standard $\alpha = 0.05$—this is the case for testC – testA. As we can see, using the plot() function on top of the TukeyHSD() function provides a quick and easy-to-understand figure with all relevant comparisons for a given variable. It is often the case that a figure communicates results more effectively than a table, so the output of line 26 is likely more effective than that of line 19 in code block 10.

2.8 More Packages

Throughout this book, we will use different packages in different chapters. More specifically, the main packages we will use are arm, brms, bayesplot, extrafont, lme4, MuMIn, ordinal, rstanarm, scales, and tidyverse. You can either install them now, all at once, by using install.packages(),[31] or install them one by one as you move through the chapters—don't worry: you will be reminded to install new packages every time. You have already installed tidyverse, which is essential, so by now you know how to do it. I will also recommend additional packages in different chapters (e.g., sjPlot, ggthemes), but you won't have to install them to reproduce the code in this book. Finally, from time to time you can update a package by running update.packages ("name-of-package").[32]

2.9 Additional Readings on R

For more resources on R, please see Wickham and Grolemund (2016). You will certainly rely on Google as you learn how to use R, so you should know some websites that can be very helpful. These are some websites that you will inevitably visit once you google different questions about R: R-bloggers, RStudio blog, Reddit (R programming language), r4stats, and Stack Overflow (r tag). Other great sources include Baayen (2008), Gries (2013), Levshina (2015), Sonderegger et al. (2018), and Winter (2019), all of which combine statistics, R, and linguistics more generally.

2.10 Summary

In this chapter, we covered *a lot*. We also created multiple scripts. First, we created rBasics.R, where we explored some general functions in R. We then created an R Project (basics.Rproj) as well as the following scripts: dataImport.R, dataPrep.R, eda.R, and stats.R. These different scripts simulate the common data analysis steps in a typical study. Always go back to Appendix D if you need to review which files have been created, their location, and where code blocks should be placed.

Following is a summary of what we discussed in this chapter—you should come back here every now and then in case you need to review the basics. As already mentioned, you are not expected to remember all the details discussed earlier.

- You can find important RStudio keyboard shortcuts in Appendix B.
- Data frames and tibbles in R are equivalent to spreadsheets in Excel. Different columns have different variables, and each row is an observation. Every column must have the same number of rows.
- Each column in a data frame or tibble is a vector, and vectors can only have objects that belong to a single class. Thus, we can have a column that only contains numbers and another column that only contains words (strings), but not a column that has both classes at the same time (numeric and character, respectively).
- Tibbles don't require head(), as they will print only the top ten rows by default. Columns that don't fit the width of your console will be omitted, but you can use glimpse() to transpose your tibble and see all columns as rows (including their individual classes).
- To change the class of a column, you can use the mutate() function. For example, to change the class of test in long from chr to Factor, you can run long = mutate(test = as.factor(test))—we will see a faster way to change all chr columns in a data frame into Factor in chapter 7.
- To access a column, you can either use a $ or slice notation. In our df data frame, we can access the second column (Languages) by typing df$Languages, df[,2], or df[, "Languages"].
- Whenever you create a variable, make sure you give it a meaningful and intuitive name (as opposed to x or y).
- Remember that variables are *not* files—your script is a file. They are not objects that are saved on your computer. Your script is saved, and you can recreate the variables by rerunning the lines in your script that generate them. You can also export variables (e.g., a data frame) as a file (.csv, .txt, etc.), in which case the file will exist on your computer beyond your script.
- Always add comments to your script so that you (and others) can clearly understand and follow the steps in your analysis.
- tidyverse contains packages that allow us to read, summarize, manipulate, and plot data.
- One of the great advantages of tidyverse is the use of intuitive functions (verbs) that perform a wide range of tasks. Important functions include group_by(), filter(), mutate(), and summarize(). Other key functions are pivot_to_long() and its counterpart, pivot_to_wide(), which allow us to transform our dataset and prepare it for analysis.
- ggplot2 is a comprehensive package for plotting data in R. It comes with tidyverse, so you don't need to install/load it as long as you install/load tidyverse.

- To build plots in ggplot2, we use "layers", which are separated by "+". The order of your layers matters, as one layer goes on top of the previous layer.
- Each layer in ggplot2 can have a different function: from plot styles to formatting adjustments—we will spend a lot of time exploring both types of functions in Part II, for example.
- You can change *anything* (every single detail) in your plot, and we will explore more specific options later on.

2.11 Exercises

It's a good idea to create a new script for the exercises here, and to do the same for the exercises in later chapters. This will make it more convenient to access them in the future.

Ex. 2.1 Vectors, Lists, and Variables

1. Problematic code A has a number of issues. Try running lines 2, 5, and 8 in RStudio and fix the code.
2. Each column in a data frame (or tibble) is a vector. In long, extract the standard deviation and median of score. Next, rename the score column to Score (uppercase S). There are different ways to do that, but let's try to use tidyverse. For that, you will need to google the rename() function.

Ex. 2.2 Using Tidyverse

1. Create a subset of long that only includes scores from testA and testB—you will need to use the filter() function in addition to the %in% operator (or the inequality operator, !=). Then use the range() function to check the lowest and highest score of the subset. Next, create a subset of long that only includes subjects whose score is greater than 7.0. For this question, you will need to source dataPrep.R, as we'll be working on long. You should do that even if long is already in your environment. This will allow the script for this exercise to be self-contained.

R code

```
1 # Creating a variable to hold five languages:
2 lgs = c(Spanish, English, French, German, Italian)
3
4 # Access second entry in lgs:
5 lgs(2)
6
7 # Create new variable for Romance languages in lgs:
8 rom = lgs[1,3,5]
```

PROBLEMATIC CODE A

2. Add a new column to long (using mutate()). The column will contain the scores of all three tests in question, but let's divide each score by 10 so we end up with a column where scores range from 0 to 1 (instead of 0 to 10). Call the new column scoreProp.
3. Using the ch2 dataset, create a summary using the summarize() function where you have the mean and standard deviation for each of the three tests by group. Do you need a wide or a long table here?

Ex 2.3 Basic Stats

1. Create a new column in ch2 that measures the improvement between testA and testB—see AtoC column created in this chapter. Then run a two-tailed *t*-test to see whether the two groups in question are different. You should find $p = 0.4515$.
2. Using the select() function, remove the comparison columns in ch2 (i.e., AtoC and AtoB created earlier). Add a new column in ch2 called testD using mutate(). The content of the column should be: rnorm(n = 10, mean = 6). Now ch2 has four test columns. Next, make a long version of the dataset and run a one-way ANOVA on the four tests. Finally, create a *post-hoc* (Tukey) plot. Is testD different from the other tests?

Notes

1. Python, another computer language, can be faster for some tasks and has a less steep learning curve relative to R, given its more consistent syntax. However, while Python is used for data analysis these days, R was designed specifically with that goal in mind. As a result, R is arguably the most specialized tool for statistical computing.
2. This analogy may soon become less accurate given the advances in computational photography.
3. Linux repositories usually have a slightly older version of R, so I'd recommend following the instructions on R's website.
4. Because line 1 ("R Basics") has just a comment without any commands to be interpreted under it, you will have to select the line first, and then run it to force RStudio to print the comment in your console. By default, RStudio won't print a comment with no commands under it if you just hit Cmd + Enter without selecting the line first. You can select a line with your keyboard by using the arrow keys to move your cursor to the beginning or end of a line and then pressing Cmd + Shift + → or Cmd + Shift + ←, respectively.
5. This operation will return the remainder of a division—cf. integer division, which discards the remainder of a division and returns the quotient.
6. You can use Cmd + Z in RStudio's script window (pane A).
7. The line numbers in a code block are not supposed to match the line numbers in RStudio (each code block starts with line 1). Line numbers are used in code blocks to facilitate our discussion on the code itself.
8. Another similar class that relates to numbers in R, integer, is often confused with numeric—the same happens with the double class. Don't worry: R will handle these different classes automatically, and for our purposes here we do not need to differentiate them. Indeed, you can successfully use R for years without even realizing that these three separate classes exist.

9. The operator in question is especially useful when you wish to examine only a subset of the data. We will practice this in the exercises at the end of this chapter.
10. You can also run help(matrix).
11. Notice how indentations are added as we break a line after a comma (e.g., lines 1–3). This is RStudio helping us organize our code by aligning the different arguments of a function (data.frame() here). We will see this throughout the book. If you select your code and press Cmd + I, RStudio will reindent all the lines to keep everything organized for you.
12. To learn how many arguments a function accepts/requires, you can press Tab after you have entered the first argument and a comma: write.csv(df,). This will open a floating menu for you with different possible arguments for the function you are using. Not all functions will explicitly list all possible arguments, but this will help you understand what different functions can do.
13. If your region uses commas as decimal separators, you can use read.csv2()—this function assumes that your columns are separated by colons, not commas.
14. Note that we spell ls() as L-S (the first character is not a number).
15. Conversely, you can use the function remove.packages() to uninstall a package. To cite a package, you can run citation("name_of_package") in R.
16. Pressing Cmd + Shift + C will comment a line out.
17. pivot_longer() and pivot_wider() are updated versions of gather() and spread(), which accomplish the same goals.
18. In code block 8, we are not assigning some of our actions to variables. You may or may not want to assign certain actions to variables, depending on whether you wish to analyze them further in your script.
19. Approach promoted by American mathematician John Tukey (1915–2000).
20. Overall, pressing Tab will auto-complete your commands in RStudio.
21. Because this is a hypothetical study, it does not matter what these scores represent.
22. This folder will help us keep our files organized by separating figures from other files, such as R scripts.
23. To avoid pixelated figures, use *at least* 300 dpi—this resolution is what most publishers require. If you like to zoom in on your figures, 1000 dpi will be more than enough.
24. You could also specify another folder, say "~/Desktop/plot.jpg", or any other path that works for you—the path in question will save the file on the desktop of a Mac.
25. Note that every time you run the function, the set of numbers generated will be different. If you want the numbers to be replicable, however, you need to be able to generate the same numbers multiple times. To do that, use the function set.seed(x) right before you run rnorm()—where x is any number of your choice (it acts like a tag, or an ID). The same applies to sample(), which is also used here.
26. This is not a very accurate term, as in reality independent variables are rarely independent.
27. Most studies in second language research do not verify statistical assumptions—see Hu and Plonsky (2020) for a recent review. Typical solutions include the removal of outliers and the use of non-parametric tests. Better solutions would be, in order of complexity, data transformation (e.g., log-transform reaction time data), the use of robust regression (instead of removing outliers), or the use of different distributions in Bayesian models that better fit the distribution at hand (e.g., chapter 10).
28. That's why you will see degrees of freedom with decimal places in your output here. If you want an integer as your degrees of freedom, add var.equal = TRUE to your t.test() function.
29. Alternatively, using %>%, we could run TukeyHSD(twoWay, which = "test")%>%plot().

30. If you don't see all three comparisons on the *y*-axis, try clicking on Zoom in your plot window (R will adjust how much text is displayed on the axes of a plot based on the resolution of your display).
31. I recommend that you wait until chapter 10 to install brms, bayesplot, and rstanarm, as you will need to follow some specific instructions. If you create a vector with all the package names in it, you can run the function only once: install.packages(c("arm", "lme4", "MuMIn", ...)).
32. How often you need to update a package will depend on the package and on your needs. To check for package updates, click on the Packages tab in pane D and then on "Update".

PART II

Visualizing the Data

3

CONTINUOUS DATA

In this part of the book, we will expand the discussion on figures started in §2.6. We will examine different types of figures and how to actually create them using R. Everything we examine in this part will come up again in Part III, where we will statistically analyze the patterns that we see in the data. To get started with data visualization in R, our focus in this chapter will be continuous variables, which are in general easier to code than categorical variables (our focus in chapter 4).

Continuous variables are quite common in second language research. Examples include test scores, reaction times, age of acquisition, number of hours exposed to a given language (e.g., per week), and so on. It's important to note that no continuous variable is truly continuous. We know that test scores are typically bound, ranging from 0 to 100, for example. Likewise, reaction times can't be negative, so they have a lower bound. Thus, none of these variables actually go from negative infinity ($-\infty$) to positive infinity ($+\infty$). However, we assume (for practical purposes) that they are continuous enough to be treated as continuous in our statistical analyses as long as they involve a range with values that are not discrete.

Let's take *age* as an example. If your study asks participants about their age, you may end up with a continuous variable *if* their responses result in an actual range, say, from 19 to 65, where you have different participants spread across that range. However, you could be in a situation where your participants are all in the same age groups: 19, 24, or 28 years old. In that case, you'd still have a range (technically), but it would make more sense to treat "age" as a discrete (ordered) variable, not as a continuous variable *per se*, given that you only have three different ages in your pool of participants.

In this chapter, we will explore different plotting techniques to visualize continuous data (i.e., a continuous response variable) using R—more specifically, the ggplot2 package. Our focus here will be the *content* of each figure. Don't worry too much about formatting or aesthetics for now, since we'll discuss that in more detail in chapter 5.

Each section that follows will focus on a different type of plot using ggplot2. We will use the same dataset across all sections in this chapter: feedbackData.csv. This is also the dataset used in chapter 6, so you will have time to get used to the data. Our task in this chapter is *not* to statistically analyze the data—we'll do that in chapter 6. Instead, we will focus on the technical component involved in plotting data using ggplot2. Therefore, focus on the different code blocks here, and make sure to reproduce the figures yourself.

DATA FILE

We will use a csv file to make figures in R. Make sure you have downloaded feedbackData.csv so you can follow every step. This file simulates a hypothetical study on two different types of feedback, namely, explicit correction and recast. The dataset contains scores for a pre-, post-, and delayed post-test. Three language groups are examined: speakers of German, Italian, and Japanese. Other variables are also present in the data—we will explore this dataset in detail later.

At this point, before proceeding to the next sections, you should create a new folder inside bookFiles and name it plots. Next, create a new R Project inside plots—the new R Project will be called plots.Rproj and will be your second R Project (basics.Rproj was created in chapter 2). These two R Projects are in two separate directories (basics and plots, respectively), so this will keep things organized. The folder we just created, plots, will be our working directory for chapters 3, 4, and 5, given that all three focus on visualization. This file organization is shown in Fig. 3.1. A complete list with all the files used in this book can be found in Appendix D—make sure you keep track of all the necessary files by visiting the appendix every now and then.

Once you've created your new project, plots.Rproj, open it and create the first script of the project—let's call it continuousDataPlots.R. All the code blocks in the sections that follow should go in that script. Fig. 3.1 also lists two other scripts under plots, namely, categoricalDataPlots.R and optimizingPlots.R, which will be created later on, in chapters 4 and 5,

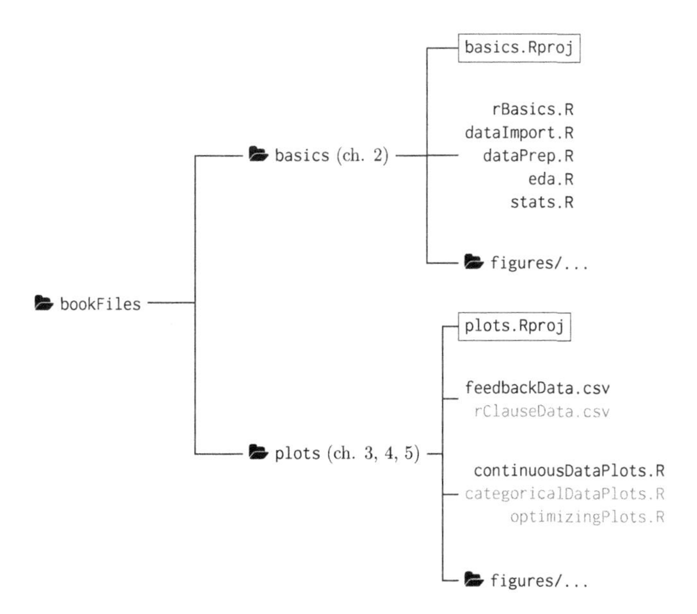

FIGURE 3.1 Organizing Your Files with Two R Projects

respectively. Lastly, we will again create a folder called figures, which is where we will save all the figures created in this part of the book—see Fig. 3.1.

3.1 Importing Your Data

Our data here represents a hypothetical study that examines the effect of different types of feedback on learning (English)—for comprehensive reviews on the role of feedback, see Russell and Spada (2006), Lyster and Saito (2010), and Loewen (2012). Our participants are learners of English and come from different linguistic backgrounds (see later description). In your new script, continuousDataPlots.R, let's first load tidyverse and then import our data file, feedbackData.csv, using the read_csv() function (see discussion later)—assign it to an easy variable to remember ("feedback"). Next, take a look at the first rows of the data (using head(feedback)). You should be able to see the following columns (variables): ID, L1, Age, AgeExp, Sex, Feedback, Hours, and ten more columns spanning from task_A1 to task_B5. You can also print a list of all variables (columns) in your data by running names(feedback). You may want to skip ahead and take a look at lines 1–18 in code block 11, but we will get there soon.

Most variables are self-explanatory: this hypothetical study divided learners of English into two groups. Both groups were taught English for a given period of time (e.g., one semester). One group received only recasts as feedback; the other group received only explicit correction. Throughout the semester, all participants were assessed ten times using two separate tasks. Their scores in each assessment can be found in the task_... columns. For example, task_B3 provides scores for assignment 3 from task B—for our purposes here, it doesn't matter what tasks A and B were. Participants' native languages were German, Italian, or Japanese. Participants also differed in terms of how many hours per week they used/studied English outside of the classroom (Hours column). Finally, AgeExp lists how old participants were when they first started studying English. Recall that if you run str(feedback), you will see the number of levels in each of the factors[1] just described (e.g., L1 is a factor with 3 levels, German, Italian, and Japanese). Importantly, you will also see that ID is a factor with 60 levels, which means we have 60 participants in the study—one row per participant in the data.

3.2 Preparing Your Data

As mentioned in chapter 2, we will rely on the concept of *tidy data* (Wickham et al. 2014) throughout this book, that is, data where every variable forms a column and each observation forms a row. Look back at the top rows of feedback—you should see 17 columns (the entire dataset has 60 rows). Is it tidy? The answer is *no*. The score of each participant is spread across *ten* different columns. For that reason, we call this type of dataset *wide*—see Table 3.1.

If feedback were tidy, we would have a single column for all the scores. Instead of ten columns for assignments and scores, we would only have three columns in this case: one identifying the task (A or B), another identifying the assignment (1–5), and a third column containing the scores (Score). This resulting dataset would be longer than the original dataset (600 *vs.* 60 rows). For that reason, we call this type of dataset *long*—see Table 3.2.

You may recall that we discussed wide-to-long and long-to-wide transformations in §2.5.1. Being able to go from wide to long to wide using R is a very important skill—we will repeat some of the steps in §2.5.1 again later. You may also recall from §2.5.1 why we have to transform feedback: if we wish to plot the scores on the x-axis of a figure, what variable should we select for the scores? Right now, we'd have to select *ten* different variables,

TABLE 3.1 Excerpt of feedback: Wide Format (60 by 17)

...	task_A1	...	task_A5	task_B1	...	task_B5
...	71.0	...	83.5	47.0	...	78.3
...	65.3	...	80.3	73.1	...	94.0
...	66.4	...	92.9	60.8	...	86.6

because scores are spread across ten columns—and this would clearly be a problem, as we'll see later. The bottom line is: you want to work with tidy (long) data.

In code block 11, we have a series of steps that mirror what we often have to do in our own research. The first line, you may recall from chapter 2, simply "cleans" the environment—this is useful if you have another script open with many variables which you don't want to use anymore (this is not the case here because this is the only script open (continuousDataPlots.R) and the only script in the directory where we also find plots.Rproj, as discussed earlier). Line 2 loads tidyverse, which itself loads ggplot2 (among other packages). Line 5 imports the data using read_csv() and assigns it to a variable (feedback). Note that read_csv() is *not* read.csv(): even though they accomplish pretty much the same task, read_csv() comes from tidyverse and creates a tibble—recall the brief discussion in chapter 2. It's really up to you if you prefer to use data frames (read.csv()) or tibbles (read_csv()). Finally, just as read_csv() can replace read.csv(), so too write_csv() can replace write.csv().

Lines 8–9 transform all characters columns into factor columns, so that different values in the columns are treated as levels; line 12 prints the top rows of the dataset,[2] line 15 prints all column names (variables), and line 18 lists the variables and their respective classes. As we'll be using this script for all the plots in this chapter, you won't have to reload the package or re-import the data (unless you close RStudio and reopen it, of course).

The key in code block 11 is lines 21–25, which are responsible for our wide-to-long transformation and which create a new variable, longFeedback. The code should be familiar, as we discussed this type of transformation when we examined code block 6.

What's new in code block 11 is line 25. If you run lines 21–24 (removing the %>% at the end of line 24), you will notice that we end up with two columns: Task and Score (defined in lines 22 and 23). The former will contain information about the task itself *and* the item: task_A1, task_A2, and so on. Line 25 *separates* this column into two columns, namely, Task and Item—in other words, line 25 separates the column we created in line 22 into two columns. The sep = 6 argument inside the separate() function specifies where we want to split the column: the sixth character from the left edge of the word. Because our Task column always contains values with seven characters (t_1 a_2 s_3 k_4 $\_{}_5$ A_6 1_7), we can easily separate this column by referring to a fixed position within the value.

Finally, line 27 visualizes longFeedback to make sure it looks good (you can also use View(longFeedback)). It should resemble Table 3.2—and it does. Remember: when we wide-to-long transform our data, all other columns remain intact. The main structural difference is that now we have a much *longer* dataset: while feedback had 60 rows (one per participant), longFeedback now has 600 rows (ten scores per participant). All the figures discussed in the next sections will rely on longFeedback.

TABLE 3.2 Excerpt of feedback: Long (Tidy) Format (600 by 10)

...	Task	Item	Score
...	task_A	1	71
⋮	⋮	⋮	⋮
...	task_A	5	83.5
...	task_B	1	47
⋮	⋮	⋮	⋮
...	task_B	5	78.3

3.3 Histograms

Histograms are useful to visualize the distribution of a given continuous variable. For example, in longFeedback, we may want to see the distribution of scores—recall that you can access that column by running longFeedback $Score. If we create a histogram of that variable, we will see that the mean

R code

```
rm(list = ls()) # To remove variables from your environment (you shouldn't have any)
library(tidyverse)

# Read file as tibble:
feedback = read_csv("feedbackData.csv")

# Transform chr to fct:
feedback = feedback %>%
  mutate_if(is.character, as.factor)

# Print top rows:
head(feedback) # or simply run "feedback"

# Print all column names:
names(feedback)

# List variables:
str(feedback)

# Wide-to-long transform:
longFeedback = feedback %>%
  pivot_longer(names_to = "Task",
               values_to = "Score",
               cols = task_A1:task_B5) %>%
  separate(col = Task, into = c("Task", "Item"), sep = 6)

head(longFeedback)

nrow(feedback)      # 60 rows before transformation
nrow(longFeedback)  # 600 rows after transformation
```

CODE BLOCK 11 Preparing our Data for Plotting (and Analysis)

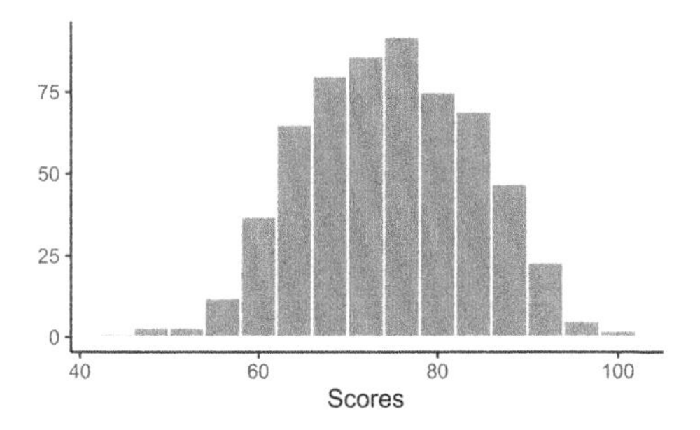

FIGURE 3.2 A Typical Histogram Using ggplot2

score is probably around 75 (naturally, running mean(longFeedback$Score) will give us an accurate answer). The histogram is shown in Fig. 3.2—if your plot looks a little different, consult Appendix A.5. Note that the *x*-axis shows the scores, and the *y*-axis shows the count, that is, the frequency of each score. We can see that most participants scored around 70 and 80 points, some scored below 60 points, and some scored above 90.

Let's now examine in detail the code that generated the histogram in Fig. 3.2—the code is shown in code block 12, and some of it will be familiar (§2.6.1). The histogram in Fig. 3.2 is generated by lines 2–9 in code block 12. You may recall from §2.6.1.2 that line 12, which is currently commented out, is used to save the figure as a jpg file. If you uncomment that line (i.e., remove the #), you can run lines 2 through 12 (simultaneously or not) and the histogram will be saved as histogram.jpg in your figures folder with great resolution (1000 dpi, as discussed in §2.6.1.2).

The first line of the histogram code (line 2) tells ggplot2 where the data comes from—this is our first layer in the figure. It also specifies what information should go on the *x*-axis and *y*-axis—note that we wouldn't be able to tell ggplot2 that our *x*-axis is Score if we hadn't transformed our data earlier. Because the *y*-axis in a histogram is typically the count of each value on the *x*-axis, we don't need to specify any variable for this axis in our code. The second layer in our plot is the histogram, which starts in line 3 and ends in line 6. The key function here is geom_histogram().

Inside geom_histogram(), we specify that we want 15 bins (bars); we want their borders to be white; we want them to have a particular width (binwidth = 4); and we want them to be partially transparent (alpha = 0.5). Lines 7–8 specify the labels of the figure (none for the *y*-axis). Finally, you may recall line 9 from code block 9—it simply adds a more minimalistic theme to the figure (no gray background, which is the default for ggplot2).

R code
```
# Histogram:
ggplot(data = longFeedback, aes(x = Score)) +
  geom_histogram(bins = 15,
                 color = "white",
                 binwidth = 4,
                 alpha = 0.5) +
  labs(x = "Scores",
       y = NULL) +
  theme_classic()

# Save plot in figures folder:
# ggsave(file = "figures/histogram.jpg", width = 4, height = 2.5, dpi = 1000)
```

CODE BLOCK 12 Producing a Histogram Using ggplot2

3.4 Scatter Plots

We may want to visually inspect how the variable Hours affects participants' scores. Intuitively, we'd expect to see a positive effect: the more participants used or studied English every week (measured in hours), the higher their scores should be on average. In that case, we'd be comparing two continuous variables, something that we can easily accomplish with a scatter plot.

Fig. 3.3 plots Score and Hours separating the data by native language (L1) into three *facets*, which are essentially panels that split our figure by a given variable. The *y*-axis represents the variable that we want to predict (Score) on the basis of other variables: we want to see how Score changes *as a function of* ("$\sim$") Hours (and L1). We can represent that relationship as y $\sim$ x + w, or Score $\sim$ Hours + L1. Here, Score is our response variable, while Hours and L1 are our predictor (or explanatory) variables.

The figure also shows a linear trend for the data points and its associated standard errors (the shaded area around the line)—this type of linear trend is the focus of chapter 6. Finally, if you inspect the axes, you will notice some internal "ticks", which display the concentration of data points on the

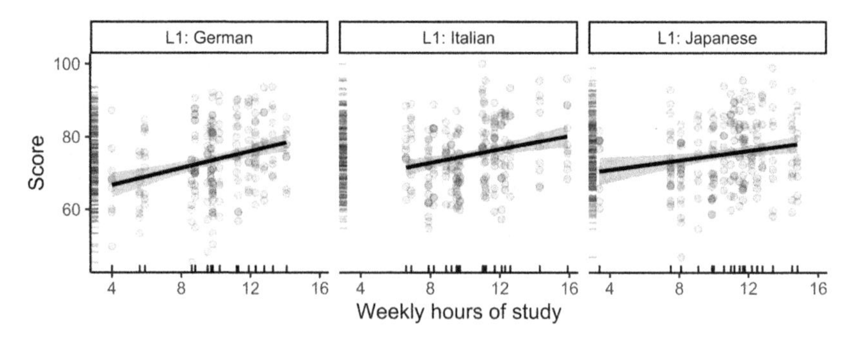

FIGURE 3.3 A Typical Scatter Plot Using ggplot2

figure—this is known as a *rug plot* layer. For German speakers, for example, we can see a higher concentration of data points between 4 and 8 weekly hours of study relative to the Japanese group.

Fig. 3.3 shows that more hours of study correlate with higher scores for all three groups—that is, a positive correlation/effect. Indeed, if you run a Pearson's product-moment correlation test in R, you will see that $\rho = 0.24$, $p < 0.001$—notice that this correlation does not distinguish the different native languages in question, so here our (micro) statistical analysis would *not* be aligned with the data visualization (cf. §2.6.2.1). You can run a correlation test in R by running cor.test(longFeedback$Score, longFeedback$Score).

Let's now inspect code block 13, which generates Fig. 3.3. Line 3, geom_point(), specifies that we want to create a scatter plot—as usual, alpha = 0.1 simply adds transparency to the data points—try removing the + at the end of line 3 and running only lines 2 and 3 to see what happens (it's useful to see what each line is doing). Line 4, stat_smooth(), is adding our trend lines using a particular method (linear model, lm, our focus in chapter 6)—the default color of the line is blue, so here we're changing it to black. Line 5 creates facets. The command facet_grid() is extremely useful, as it allows us to add one more dimension to our figure (here, native language, L1). Notice that we use a tilde (~) *before* the variable by which we want to facet our data.

As long as we have a categorical/discrete variable, we can facet our data by it. We can even add multiple facets to our figures. For example, facet_grid(A ~ B) will generate a figure with variable A as columns and variable B as rows. In longFeedback, we could facet our data by both L1 and Task—each of which has two levels. The result would be a 3-by-2 figure. Alternatively, if you prefer a wide figure (with 6 horizontal facets), you can type facet_grid (~ A + B), and if you want a tall figure (with 6 vertical facets), you can type facet_grid(A + B ~ .)—don't forget the "." on the right-hand side. The argument labeller = tells ggplot2 that we want to label our facets (i.e., we want to print the name of the factor, not just the names of the levels). In our code, we're asking ggplot2 to label both the variable (the name of the factor), for example, L1, and the values (the content of the levels), for example, German.

Line 8 in code block 13, geom_rug(), adds the marginal ticks along the axes. Notice that we can also add transparency to our rugs, by adding alpha = 0.2 to geom_rug()—you could also add color (the default is black). Whenever you want to learn what arguments a given function takes, you can run help (geom_rug) or ?geom_rug() (for the example in question).

The trends shown in Fig. 3.3 ignore the different tasks and assignments in the hypothetical study in question. In other words, we are looking at scores as a whole. But what if the relationship between Score and Hours is different depending on the task (A or B)? Let's adjust our figure to have two different trend lines based on the two levels of Task in the data. The new plot is shown in Fig. 3.4—note that we're using different line types for each task

R code
```
1  # Scatter plot:
2  ggplot(data = longFeedback, aes(x = Hours, y = Score)) +
3    geom_point(alpha = 0.1) +
4    stat_smooth(method = lm, color = "black") +
5    facet_grid(~L1, labeller = "label_both") +
6    labs(x = "Weekly hours of study",
7         y = "Score") +
8    geom_rug(alpha = 0.2) +
9    theme_classic()
10
11 # Save plot:
12 # ggsave(file = "figures/scatterPlot1.jpg", width = 6, height = 2.3, dpi = 1000)
```

CODE BLOCK 13 Producing a Scatter Plot Using ggplot2

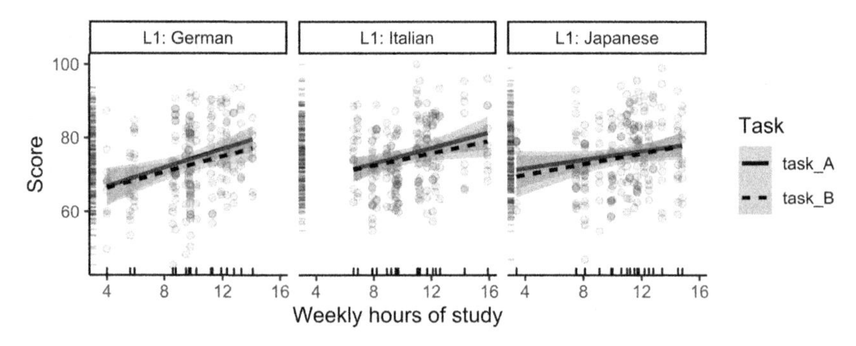

FIGURE 3.4 A Scatter Plot with Multiple Trend Lines Using ggplot2

(solid line for A, dashed line for B). As we can see, the correlation between scores and hours of study doesn't seem to be substantially affected by Task.

The code to generate Fig. 3.4 can be found in code block 14. Pay attention to line 4, where we have stat_smooth(). One of the arguments in that function is aes()—the same argument we use in our first layer to specify our axes. There we define that we want trend lines to be of different types *based on* a variable, namely, Task. Notice that we are specifying that *locally*. In other words, only stat_smooth() has the information that we want to differentiate the two tasks. The other plotting layers, such as geom_point(), will therefore disregard Task. To specify an argument *globally*, you can add it to the first layer, in line 2, inside aes() in the ggplot() function (we already do that by telling ggplot() what our axes are; the other layers simply inherit that information).

Here's an example to help you understand the effects of local and global specifications in a figure using ggplot2. Imagine that we wanted to produce Fig. 3.4, but using colors instead of line types. You could have as your first layer ggplot(data = feedback, aes(x = PreTest, y = PostTest, color

R code

```
1  # Scatter plot:
2  ggplot(data = longFeedback, aes(x = Hours, y = Score)) +
3    geom_point(alpha = 0.1) +
4    stat_smooth(aes(linetype = Task), method = lm, color = "black") +
5    facet_grid(~L1, labeller = "label_both") +
6    labs(x = "Weekly hours of study",
7         y = "Score") +
8    geom_rug(alpha = 0.2) +
9    scale_color_manual(values = c("black", "gray")) +
10   theme_classic()
11
12 # Save plot:
13 # ggsave(file = "figures/scatterPlot2.jpg", width = 6, height = 2.3, dpi = 1000)
```

CODE BLOCK 14 Producing a Scatter Plot with Multiple Trend Lines Using ggplot2

= Task)). Here you'd be defining a functional use of color *globally*. As a result, geom_point(), stat_smooth(), and geom_rug() would inherit that: you'd end up with data points in two colors and trend lines in two colors (they would match). If instead you specified color = Task only inside geom_point() and removed it from your first layer, then only the data points would display proficiency levels in two colors—and you'd no longer have two trend lines per Task.

Let's produce one more scatter plot. This time, let's have a simpler figure: no trend lines, no rug plots. Only now we want to add Age as another dimension to our figure. For example, we could make the size of each data point proportional to the age of the participant that produced said data point—this may or may not be useful depending on your data and on your objectives.

Before we inspect our new scatter plot, however, let's go back to the structure of our data. Recall that we have 60 participants and that each participant has 10 scores (hence the 600 rows in longFeedback). By definition, then, each participant has 10 points on our scatter plots—so we have 600 points in Fig. 3.4, for example.

The problem is that 600 points is *a lot* of points—especially if we want the size of each point to represent participants' ages. Besides, we don't really need 10 separate points for the same participant once we want to include Age. Think about it this way: each participant has a constant number of weekly hours of study and a constant age in our dataset. As a result, in longFeedback, we have to repeat each participant's Age and Hours ten times (one for each score). What if we calculate the *mean* score across all ten assignments for each participant? If we did that, we would have a new table with one row per participant, and we would no longer need 10 separate points on our figure to represent each individual. A scatter plot with 60 points is certainly easier to visually inspect than one with 600 points—especially when the size of each point represents a continuous variable (Age). Our new plot is shown in Fig. 3.5.

Fig. 3.5 is essentially plotting three continuous variables (mean Score, Hours, Age) and separating them based on a fourth discrete variable using facets (L1). As it turns out, Age doesn't seem to affect the other variables in Fig. 3.5—which is not very surprising. The different point sizes used to represent different ages can be added to our plot by adding aes(size = Age) to geom_point() (local specification) or by adding it to the first layer, which is what code block 15 shows. Clearly, this figure is not very informative, mostly because Age doesn't seem to make a difference here: we see larger circles at the top *and* at the bottom.

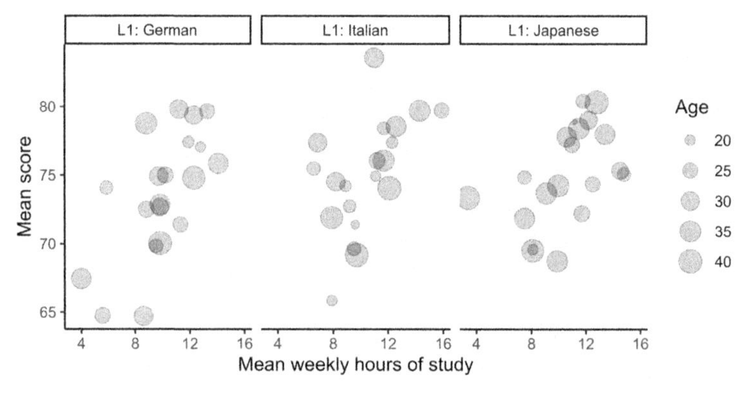

FIGURE 3.5 A Scatter Plot with Point Size Representing a Continuous Variable

R code

```
# Group and summarize data:
ageSummary = longFeedback %>%
  group_by(ID, L1, Hours, Age) %>%
  summarize(meanScore = mean(Score))

head(ageSummary)

# Scatter plot: representing Age with size
ggplot(data = ageSummary, aes(x = Hours, y = meanScore, size = Age)) +
  geom_point(alpha = 0.2) +
  facet_grid(~L1, labeller = "label_both") +
  labs(x = "Mean weekly hours of study",
       y = "Mean score") +
  scale_color_manual(values = c("black", "gray")) +
  theme_classic()

# Save plot:
# ggsave(file = "figures/scatterPlot3.jpg", width = 6, height = 3, dpi = 1000)
```

CODE BLOCK 15 Producing a Scatter Plot with Three Continuous Variables Using ggplot2.

Code block 15 shows how Fig. 3.5 is generated as well as how the data to be plotted is prepared (summarized). In lines 2–4, we create a new variable, ageSummary, which groups our data by four variables, ID, L1, Hours, and Age—line 3. Line 4 then summarizes the data by creating a column that calculates the mean Score. Note that *all* four variables are constant in the data: Learner_1 always has the same value for L1 (Japanese), Hours (8.1), and Age (38). In other words, there's no nesting involved in our grouping: we're simply grouping by all four variables because we want our resulting tibble (ageSummary) to have these variables as columns, otherwise we wouldn't be able to refer to these variables in our figure—run line 6 to visualize the top rows of ageSummary to make sure you understand what lines 2–4 are doing. We will discuss data summarization again in chapter 4.

The take-home message here is that you can present different variables using different layers in your figure. Layers can display multiple dimensions of data using shapes, colors, line types, sizes, facets, and so on. Naturally, which options will work depends on what your figure looks like: if you're creating a scatter plot with no trend lines, using linetype to represent a given variable will be semantically vacuous.

Finally, take your time to inspect code blocks 13, 14, and 15. The fact that they look very similar is good news: ggplot2 always works the same way, layer by layer, so it's a consistent package in terms of its syntax. In the next sections, we will examine other useful plots for continuous data, and you'll have the chance to see more layers using ggplot2—they will again look similar.

3.5 Box Plots

Box plots are underrated and underused in second language research. This is really unfortunate, given how useful they are. Assuming that your data is roughly normally distributed, box plots are reliable and very informative as long as you have a discrete or categorical variable on your *x*-axis and a continuous variable on your *y*-axis. A box plot typically contains three elements: a box, whiskers, and outliers. Let's briefly examine all three in Fig. 3.6.

The gray box in Fig. 3.6 represents all the data from the first quartile (Q1, 25th percentile) to the third quartile (Q3, 75th percentile)—that is our

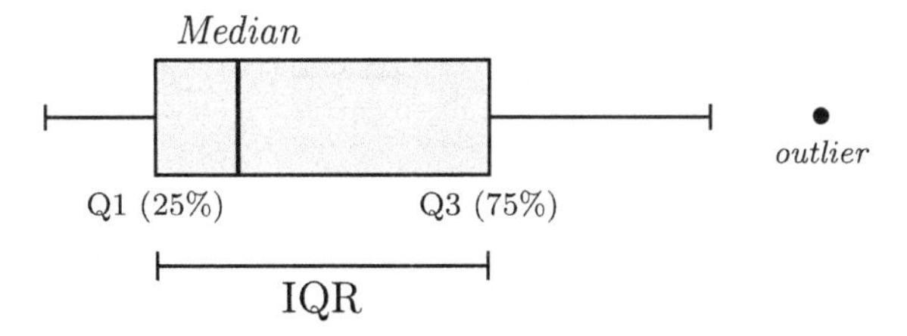

FIGURE 3.6 The Structure of a Box Plot

interquartile range (IQR). Simply put, the box represents the 50% most frequent data points in any given dataset, that is, the "bulk" of the data. Within the box, we can see a vertical line representing the median (second quartile, or 50th percentile), so we know that 50% of the data is below the line and 50% of the data is above the line. The whiskers represent the remainder of the data *excluding outliers*. Technically speaking, the lower whisker (on the left) extends to Q1 −1.5 · IQR, whereas the upper whisker (on the right) extends to Q3 + 1.5 · IQR. Thus, the box and the whiskers account for 99.3% of the data. You may not have outliers, of course (Fig. 3.7 doesn't), but if you do, they will be displayed as individual points in your box plot.

You don't have to remember the exact percentage points involved in a box plot. Just remember that a box plot shows the spread of the data as well as its median, so it provides a very informative picture of our datasets. Focus on the box first, and then check how far the whiskers go.

Let's now inspect Fig. 3.7. Here we have a 2-by-3 plot, and we have four different dimensions (variables): on the *y*-axis we have Score, a continuous variable; on the *x*-axis we have Feedback, a factor with two levels; and then we have two sets of facets, one for L1 (columns) and another for Task (rows). In addition to box plots, note that we also have semitransparent data points on the background, so the reader has access to the actual data points as well as the box plots showing the spread of the data.

Overall, we can see in Fig. 3.7 that Recast seems to yield higher scores relative to Explicit correction. As a rule of thumb, the more two box plots overlap, the less likely it is that they are statistically distinct. For example, if you inspect the facet for Japanese speakers, you will notice that box plots for Explicit correction and Recast overlap almost completely.[3] As a result, these

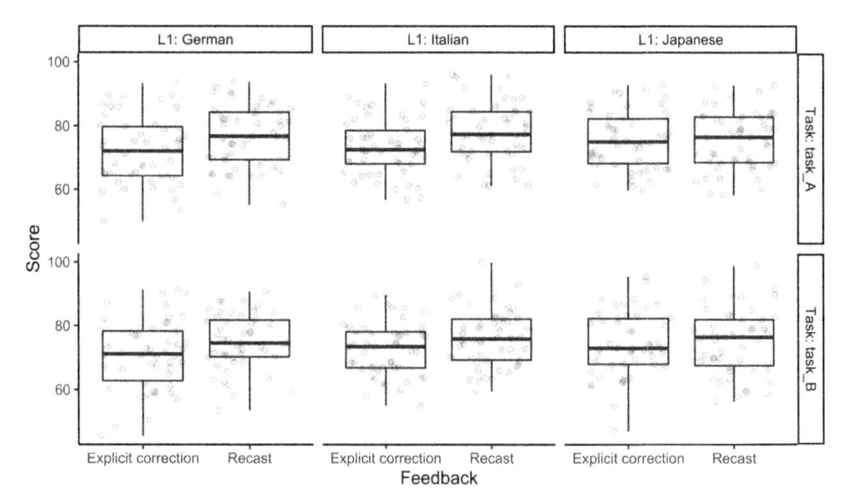

FIGURE 3.7 A Box Plot (with Data Points) Using ggplot2

learners don't appear to be affected by Feedback the same way as the other two groups in the figure.

Code block 16 has some new lines of code. First, in line 3, we have geom_jitter() with some transparency. This function spreads data points that would otherwise be stacked on top of each other. Look back at our *x*-axis. Notice that we have a *discrete* variable there. As a result, all the data points would be vertically aligned—which wouldn't make for a nice figure to look at. By using geom_jitter(), we ask ggplot2 to spread those data points.

Next, in line 4, we have our main function: geom_boxplot(). Here, we're using alpha = 0, which means the box plots will be completely transparent (they're gray by default). Remember once more: to create a box plot, you must have a discrete or categorical variable on your *x*-axis and a continuous variable on your *y*-axis. We then use facet_grid() in line 5 to create our two facets—at this point you should be familiar with lines 6 and 7 as well as line 10. Finally, note that we're not specifying the label for the *x*-axis here because the name of the variable is already appropriate (Feedback). ggplot2 will label the axes by default with their respective variable names, so we only need to specify our own labels if we're not happy with the default labels—this was the case with Hours in previous plots, which wasn't self-explanatory. Likewise, ggplot2 will automatically order the levels of a factor alphabetically along an axis. As a result, Explicit correction is shown before Recast. We could, however, reorder the levels based on Score in this particular case—this is especially useful when several levels are present on the *x*-axis. To do that, we would use the fct_reorder() function inside aes(): ..., aes(x = fct_reorder(Feedback, Score), y = Score).

3.6 Bar Plots and Error Bars

Bar plots are one of the most common figures in data analysis. They are typically used to represent a continuous variable on the *y*-axis and a categorical

R code

```
1  # Boxplot + jitter + facets:
2  ggplot(data = longFeedback, aes(x = Feedback, y = Score)) +
3    geom_jitter(alpha = 0.1) +
4    geom_boxplot(alpha = 0) +
5    facet_grid(Task ~ L1, labeller = "label_both") +
6    theme_classic() +
7    labs(y = "Score")
8
9  # Save plot:
10 # ggsave(file = "figures/boxPlot.jpg", width = 7, height = 4, dpi = 1000)
```

CODE BLOCK 16 Producing a Box Plot Using ggplot2

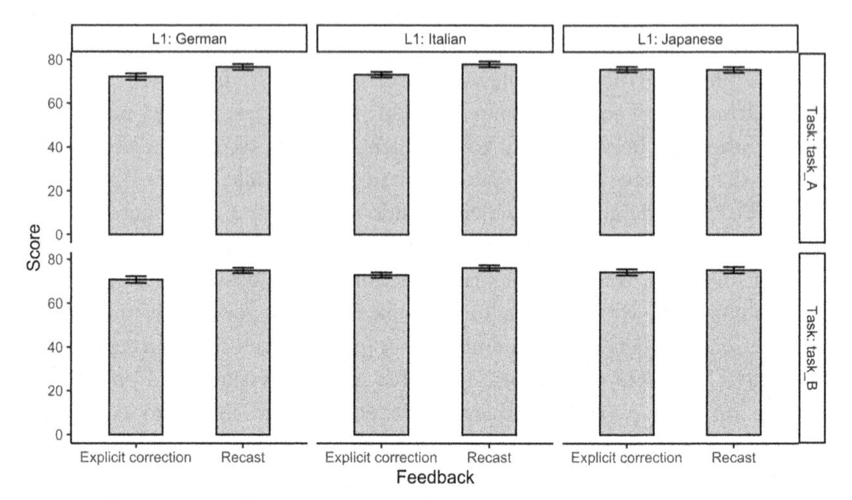

FIGURE 3.8 A Bar Plot (with Error Bars) Using ggplot2

variable on the *x*-axis. Fig. 3.8, for example, plots the mean Score by Feedback and L1. In addition to mean values, Fig. 3.8 also displays the standard error from the mean using error bars. Error bars showing standard errors are very important, as they allow us to estimate our level of uncertainty around the means (see §1.3.4). You should not have bar plots without standard errors when you show variable (experimental) data: the bars themselves don't provide a lot of information (only the means in this case).

You may have noticed that the bar plot in Fig. 3.8 is structurally equivalent to the box plot in Fig. 3.7, since both figures plot the same variables. But what differences do you notice? Box plots show us the spread of the data as well as the median. Bar plots (and error bars) show us the mean and its standard error. Look carefully at the standard error bars in Fig. 3.8, which are very small. You will notice that there's very little overlap between explicit instruction and recast. In contrast, our box plots in Fig. 3.7 show massive overlap. This difference makes intuitive sense if you know what each figure is showing. But assume that your readers don't understand box plots that well and that you are not showing them bar plots with error bars—your paper only has box plots. Readers may conclude that there's likely no effect of Feedback based on the box plots. If you had shown them the bar plots, on the other hand, they would likely conclude that Feedback has an effect, given that the error bars don't overlap in most facets in Fig. 3.8.

The take-home message here is this: these plots are showing the same data, and the patterns they display are *not* contradictory. They are simply two different perspectives of the same variables and effects. Finally, bear in mind that Fig. 3.8 is not aesthetically optimal for a couple of reasons. First, it's quite large, even though there's a lot of empty space in the figure—we only have two bars per facet. We could certainly improve it by removing Task from a facet and using different bar colors (fill) to represent them (e.g., two shades of gray). In that case, we'd end up with four bars per facet, and our figure would only have one row of facets (as opposed to two). Second, the standard errors are very small, which makes it hard to inspect potential overlaps. We could start our *y*-axis from 60, which would show us a "zoomed" version of the figure. That may not be optimal, however, since it tends to overemphasize small differences. Alternatively, we could remove the bars and keep only the error bars (which would automatically adjust the *y*-axis to only include the relevant range for the standard errors). Ultimately, a bar plot is probably not the best option here (aesthetically speaking).

The code provided in code block 17 should be familiar by now (recall the first plot we examined back in §2.6.1). The two main layers of the figure in question are the bars (line 3) and the error bars (line 4). Both layers are the result of stat_summary(), a very useful function that combines plotting and some basic statistics. Here, we are not manually calculating means or standard errors: both are calculated on the go by stat_summary() as we plot the data. If we flipped the order of these layers, the error bars would be behind the bars—which wouldn't be a problem here because our bars are completely transparent. Finally, you could use error bars to represent the standard deviation in the data instead of the standard error. You can do that by replacing line 4 with stat_summary(fun.data = mean_sdl, geom = "errorbar", width = 0.2). When we display the standard error, we are focusing on our uncertainty about the sample mean; when we display the standard deviation, we are focusing on the variation of the actual data being plotted.

You may be wondering: is it possible to combine the level of information of a box plot with the mean and standard error of a bar plot with error bars? The answer is *yes*. The great thing about ggplot2 (well, R in general) is that we have total control over what happens to our data and figures. Fig. 3.9 combines a box plot with two different error bars: solid error bars represent the SE from the mean, and dashed error bars represent *s* in the data.[4] The great thing

R code

```
1  # Bar plot:
2  ggplot(data = longFeedback, aes(x = Feedback, y = Score)) +
3    stat_summary(geom = "bar", alpha = 0.3, color = "black", width = 0.5) +
4    stat_summary(geom = "errorbar", width = 0.2) +
5    facet_grid(Task ~ L1, labeller = "label_both") +
6    theme_classic() +
7    labs(y = "Score")
8
9  # Save plot:
10 # ggsave(file = "figures/barPlot.jpg", width = 7, height = 4, dpi = 1000)
```

CODE BLOCK 17 Producing a Bar Plot Using ggplot2

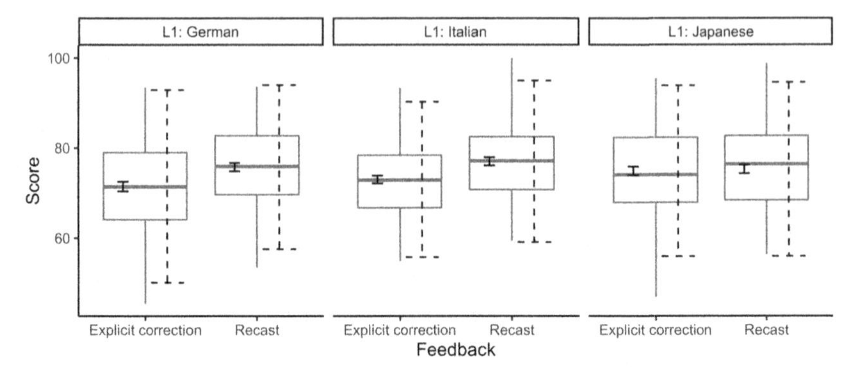

FIGURE 3.9 A Box Plot with Error Bars for Standard Errors and ±2 Standard Deviations

about such a plot is that it gives us a lot of information about the data. Is it overkill? Probably, so you may want to comment out lines 6–8 and only show box plots and SEs from the mean. Box plots that also display means and standard errors are often everything you need to see about your data.

The code for Fig. 3.9 can be found in code block 18. The layers of the figure are straightforward: we have one layer for the box plot (line 3), one layer for the standard error bars (lines 4–5), and one layer for the standard deviation bars (lines 6–8)—this time our box plot is gray, so it acts like a faded background image. Notice that `position` is added to both error bars such that SEs move left (`position` = `–0.2`) and *s* move right (`position` = `0.2`). If you don't do that, both error bars will be centered, which means they will be positioned on top of each other. Also notice the `linetype` argument in line 8 to make the *s* error bars dashed.

The nice thing about Fig. 3.9 is that it shows us that although the box plots overlap quite a bit, the means and standard errors do not (except for the Japanese speakers). Overall, the figure also shows us that the means and the medians in the data are very similar (common with data points that are normally distributed).

In summary, bar plots are simple but effective if you want to focus on means. If you combine them with error bars and facets, they can certainly be informative. We also saw earlier that error bars can certainly be combined with box plots. The key is to generate figures that hit the sweet spot in terms of information and clarity.

3.7 Line Plots

Line plots are used to show how a particular continuous variable *y* changes as a function of another (continuous) variable *x*. Ideally, both axes on your figure should be continuous, since a line implies continuity between points.[5] Let's

R code

```
1  # Box plot + SE + SD:
2  ggplot(data = longFeedback, aes(x = Feedback, y = Score)) +
3    geom_boxplot(alpha = 0, color = "gray50") +
4    stat_summary(geom = "errorbar", width = 0.1,
5                 position = position_nudge(x = -0.2)) +
6    stat_summary(fun.data = mean_sdl, geom = "errorbar", width = 0.3,
7                 position = position_nudge(x = +0.2),
8                 linetype = "dashed") +
9    facet_grid(~L1, labeller = "label_both") +
10   theme_classic() +
11   labs(y = "Score")
12
13 # Save plot:
14 # ggsave(file = "figures/boxSeSdPlot.jpg", width = 7, height = 3, dpi = 1000)
```

CODE BLOCK 18 Adding Error Bars to a Box Plot

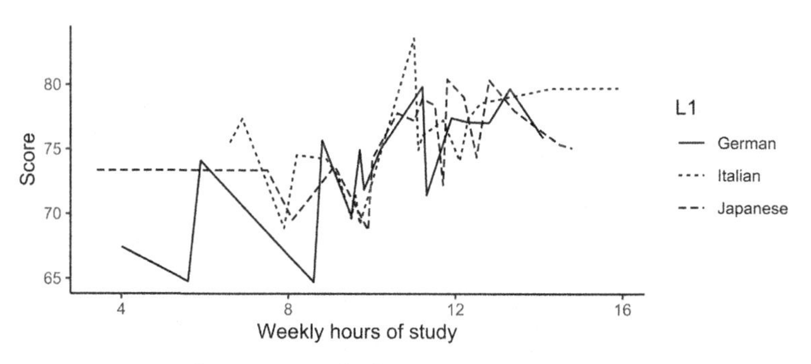

FIGURE 3.10 A Line Plot Using ggplot2

examine the example in Fig. 3.10, where we have scores on the *y*-axis and weekly hours of study (Hours) on the *x*-axis.

Line plots *resemble* scatter plots structurally, since both axes are continuous. In Fig. 3.10, we see three lines, one by group (L1). The lines represent the mean scores of each group as a function of the number of weekly hours of study. We see a similar trend for all groups: on average, more weekly hours lead to higher scores.

If you examine code block 19, you will notice that the function stat_summary() is again the key to our figure. In line 3, we specify that we want to plot means with lines by adding geom = "line". The third step is to add aes(linetype = L1), which by now you should be familiar with: this uses different types of lines depending on the value of the variable L1—recall that this variable must be a factor. Because the factor in question has three levels (German, Italian, and Japanese), three lines are plotted.

R code

```
# Line plot:
ggplot(data = longFeedback, aes(x = Hours, y = Score)) +
  stat_summary(geom = "line", aes(linetype = L1)) +
  labs(y = "Score", x = "Weekly hours of study") +
  theme_classic()

# Save plot:
# ggsave(file = "figures/linePlot.jpg", width = 6, height = 2.5, dpi = 1000)
```

CODE BLOCK 19 Producing a Line Plot Using ggplot2

Line plots can be especially useful if you collect longitudinal data. However, as we can see in Fig. 3.10, they can also be useful to show trends across a time variable (most of us collect extra-linguistic information that falls into that category). Naturally, if you're looking at ten different languages, a line plot may not be appropriate (too many lines). But you should create the figure and examine the trends nonetheless. Whether the figure will be in your actual paper is a completely different story.

3.8 Additional Readings on Data Visualization

We will discuss data visualization at length in this book—it is the focus of Part II, and Part III will discuss visualization again. However, our discussion will not be exhaustive, naturally. If you wish to study more advanced features of ggplot2, for example, see Wickham (2016). You will most likely use Google while producing figures in R: forums on websites such as Stack Overflow can be incredibly useful (see questions tagged with ggplot2 here: https://stackoverflow.com/questions/tagged/ggplot2). Likewise, you may wish to bookmark ggplot2's official website: https://ggplot2.tidyverse.org/.

3.9 Summary

In this chapter, we discussed some general guidelines for visualizing data and examined the most common plots for continuous variables using ggplot2. All the figures discussed earlier were generated by using different layers of code and should be saved in the figures folder (to keep scripts and figures separated within an R Project). The first layer identifies which data object should be plotted and specifies the axes. Once the first layer is defined, we can use different layers to create and adjust our figures. Here is a summary of the main functions we used.

- Histograms: geom_histogram(). Histograms plot the distribution of a given continuous variable. Here, your first ggplot2 layer doesn't need to specify the *y*-axis. Refer to code block 12.

- Scatter plots: geom_point(). Scatter plots are used when you have continuous variables on both axes in your figure. These plots are ideal for displaying correlations. Refer to code block 13.
- Trend lines: stat_smooth(method = lm). Trend lines can be added to a figure (e.g., scatter plot) by adding the function stat_smooth() (or geom_smooth()) to our code. Refer to code block 13.
- Box plots: geom_boxplot(). Box plots are used when your x-axis is categorical or discrete. These plots are very informative as they show the spread of the data as well as the median. Refer to code block 16. We also saw how to include means and error bars on box plots—refer to code block 18.
- Bar plots: stat_summary(). Like box plots, bar plots are used when your x-axis is categorical or discrete. Even though there is a function called geom_bar(), you can generate bar plots using stat_summary(). You should add error bars to your bar plot whenever it's appropriate. Refer to code block 17.
- Line plots: stat_summary(). Line plots are useful to show how a given continuous variable (y-axis) changes as a function of another (ideally) continuous variable (x-axis). Refer to code block 19.
- Facets: facet_grid(). Facets in ggplot2 allow us to add more variables to our figures, which will be plotted on different panels. Adding a facet to our bar plot in Fig. 3.8 would have made it more informative (e.g., facet_grid (~proficiency)). Refer to code block 14, for example.
- aes() allows us to specify the variables that will be displayed along the y- and x-axes. Crucially, it also allows us to specify how we wish to plot additional dimensions. For example, if we wish to plot L1 using bars with different colors (each color representing a proficiency level), we would add aes(..., fill = L1) to our code—of course, if you do that you shouldn't facet by L1. If we add that line of code to our first layer, its scope will be global, and all other layers will inherit that specification. If we instead have a line plot, we must replace fill with color, since lines have no fill argument. Thus, by enriching the argument structure of aes() we can add more dimensions to our plots without using facets (you can naturally use both). Refer to code block 19, for example, where we use different line types for the different languages in our dataset.

3.10 Exercises

Ex. 3.1 Plots in R

1. In feedbackData.csv, each task contains five assignments. Create a figure to answer the following question: do participants' scores change across assignments for the two tasks in our hypothetical study? The first step is to note that the x-axis will have *categories* (assignments) and the y-axis will have scores, which are continuous. This is a very common scenario, which

R code

```
1 ggplot(data = longFeedback %>%
2          filter(Age > 40, Task = "task_A"),
3        aes(x = L1, y = Score)) +
4   stat_summary() +
5   facet_grid(~Task)
```

PROBLEMATIC CODE B

is compatible with bar plots, box plots, error bars, and so on. Try answering this question with box plots and different fill colors for Task. Then try again with bars (and error bars) to see how they compare.

2. Using longFeedback, create a histogram showing the distribution of Score—see Fig. 3.2. This time, however, add fill = Feedback to aes() in the first layer of your plot. The resulting plot will have *two* histograms overlaid in a single facet.

Ex. 3.2 Filtering Data Inside Ggplot()

1. We often want to plot parts of our data, but not the entire dataset. Take our line plot in Fig. 3.10: we might want to remove one of the L1 groups from our data to have a plot with only two groups. We have already discussed how we can subset our data using the filter() function (Table 2.3). The nice thing is that you can use that function *inside* ggplot(). Go ahead and create Fig. 3.10 without the Japanese group. Simply add %>% after the data variable in the first layer of the plot in code block 19—indeed, you could do a lot of complicated operations using %>% inside ggplot().
2. Reproduce Fig. 3.3, but this time only plot scores greater than 60 and Hours greater than 8. You can add multiple filters either by separating the conditions with a comma or "&". *Hint:* If you feel stuck, do the next question first and then come back.
3. Observe problematic code B. RStudio will try to help you with syntax errors when you run the code, but it won't tell you much more than that. Your task is to make the code work: it should generate a plot with mean scores by group (and associated error bars). *Hint:* You may have to inspect the data to figure out some of the issues in the code.

Notes

1. Once you run str(feedback), you will likely see that some variables are character variables (chr), not factors (fct). To see them as factors, run lines 8–9 from code block 11.
2. Because feedback is a tibble, we could just run feedback here, without head().
3. Remember, however, that box plots do not show us the *mean* scores of the groups, which is what we will focus on in our statistical analysis in chapter 6.

4. By default, mean_sdl plots ±2 standard deviations from the mean. You can change that by adding fun.args = list(mult = 1) to stat_summary() (see lines 6–8 in code block 18).
5. You will often see line plots where the x-axis is categorical, in which case lines are used to connect discrete points and to convey some trend. Whether or not this is recommended depends on a number of factors, including the nature of your data and what you want to show.

4

CATEGORICAL DATA

Consider the following situation: your participants complete a survey where they can respond yes or no to several questions, so you have a *binary* response variable. In such a scenario, we typically plot counts or percentages of each response. For example, "40% of the responses were yes". But your dataset doesn't have that information. As a result, there isn't a variable in your data to occupy the y slot in your first plot layer: ggplot(data = ..., aes(y = ?)). Here, we have to compute these percentages ourselves.[1]

In chapter 3, our response was a test score, that is, a continuous variable. That allowed us to simply specify y = PreTest in our code for box plots, for example. When we deal with a categorical response variable, however, things get a bit more complicated, and we will most likely have to first prepare and transform our variable(s) before making plots.

In this chapter, we will see how to transform categorical variables into continuous variables (counts or percentages). Once we have a continuous variable, the skills you learned in chapter 3 can be easily transferred. Later, we first examine binary data, where responses are essentially 1 or 0 (as in the yes-no example earlier). We will then consider ordinal data, where our variable of interest is a numeric scale. Before we start, let's briefly examine the dataset we will use for this chapter as well as chapter 5—don't create any scripts just yet; we will do that shortly.

DATA FILE

Make sure you have downloaded rClauseData.csv. This file simulates a hypothetical study on relative clauses in second language English.

Examining Our Data

The data we will use involves a hypothetical study about relative clauses in English as a second language. Take a look at the sentence in (1a). Because cats don't dance (at least not intentionally), we conclude that the relative clause refers to the noun phrase (NP) *the nurse*. But interpretations are not categorical when a semantic bias is not available. For example, in sentence (1b), either the NP *the daughter* or *the nurse* can be the head for the relative clause *who likes to dance*.

(1) **Relative clauses: high (NP_1) and low (NP_2) attachment**

a. Mary saw [the cat]$_{NP_1}$ of [the nurse]$_{NP_2}$ who likes to dance
b. Mary saw [the daughter]$_{NP_1}$ of [the nurse]$_{NP_2}$ who likes to dance

This type of syntactic ambiguity has been well studied across languages (e.g., Cuetos and Mitchell 1988, Fodor 2002, Fernández 2002, Goad et al. 2020), and different languages have been shown to favor different interpretations in such cases. English, for example, tends to favor LOW attachment, that is, NP_2. Spanish, in contrast, has been shown to favor HIGH attachment, that is, NP_1. Therefore, while an English speaker would be more likely to assume that it's the nurse who likes to dance in (1b), a Spanish speaker would be more likely to assume that it's the daughter.

Our data comes from a hypothetical auditory experiment comparing speakers of English (controls) and speakers of Spanish learning English. Participants heard different sentences that were ambiguous (e.g., (1b)) and were subsequently asked "who likes to dance?" Crucially, the experiment also investigates the possible role of a pause, given that prosody has been shown to affect speakers' interpretations (Fodor 2002). In (2), we see all three relevant conditions ([#] = pause).

(2) **Three conditions in target items**

a. Mary saw the daughter of the nurse who likes to dance	NoBreak
b. Mary saw the daughter [#] of the nurse who likes to dance	High
c. Mary saw the daughter of the nurse [#] who likes to dance	Low

Sentences were recorded which contained no pause (NoBreak), a pause after NP_1, and a pause after NP_2. For condition (2a), we expect native speakers of English to favor low attachment and learners to favor high attachment (transferring the pattern from their native language, Spanish). The question is what happens when a pause is present, that is, whether our two different groups behave differently, whether proficiency matters, and so on. We will start addressing these questions later as we generate some figures to visualize the patterns in the data. Later, in chapter 7, we will work on our statistical analysis.

Getting Started

Create a new script called categoricalDataPlots.R. Because our topic of discussion is still data visualization, you can save your script in the same directory used in chapter 3—we will also use the figures folder created in the previous chapter, since we're still in the same R Project. Thus, the scripts from chapters 3 and 4 (and 5) will belong in the same R Project, namely, plots.Rproj—refer to Fig. 3.1. Finally, in categoricalDataPlots.R, load tidyverse, and import rClauseData.csv (you can use either read.csv() or read_csv())—assign it to variable rc (*r*elative *c*lause).

If you visualize the top rows of rc using head(rc), you will see that our dataset has 12 variables: ID (participants' ID); L1 (participants' native language, English or Spanish); Item (each sentence used in the experiment); Age (participants' age); AgeExp (learners' age of exposure to English); Hours (learners' weekly hours of English use); Proficiency (participants' proficiency: Int or Adv for learners or Nat for native speakers of English); Type (Filler or Target items); Condition (presence of a pause: High, Low, or NoBreak); Certainty (a 6-point scale asking participants how certain they are about their responses); RT (reaction time); and Response (participants' choice: High or Low). There are numerous NAs in rc. This is because native speakers are coded as NA for AgeExp and Hours, for example. In addition, Condition and Response are coded as NA for all the fillers[2] in the data (given that we only care about target items). So don't worry when you see NAs.

Some of the variables in rc are binary, namely, L1, Type and, crucially, Response—our focus later (§4.1). To examine how we can prepare our data and plot some patterns, we will examine how Proficiency and Condition may affect speakers' responses—clearly the focus of the hypothetical study in question.

4.1 Binary Data

If you try to plot participants' responses as a function of their proficiency level using ggplot2, you will get an odd-looking figure. Try this: ggplot(data = rc, aes(y = Response, x = Proficiency))+ stat_summary(). The result is a figure with three levels on the *x*-axis (Adv, Int, Nat).[3] The figure also has three values on the *y*-axis: High, Low, and NA. This is because fillers are included, and responses for fillers are coded as NAs, as mentioned earlier. Needless to say, this figure is useless: it simply displays nine equally spaced data points that tell us nothing about the data. Because our *y*-axis is discrete, not continuous, stat_summary() can't compute standard errors unless we give it a numeric variable, of course.

What we need to know here is *how often* participants choose a given response for each condition of interest (and for each proficiency level in question). This can be accomplished if we count how many High and Low responses we have per condition and proficiency level. Of course, if our data is not balanced, counts will be misleading, so we should consider using percentages instead.

Fig. 4.1 is what we're looking for. Let's examine it first and then discuss how you can create the figure yourself. On the *y*-axis, we have the percentage of low attachment responses (technically, the axis shows decimals and the values don't go all the way up to 1, but we will adjust that in chapter 5). Because our response variable is binary, we can pick high or low attachment as our reference here. We're choosing low attachment because that's the default choice for English, and this hypothetical study is about acquiring English as a second language. The *x*-axis shows the different proficiency levels: Adv, Int, Nat. Finally, the different facets separate the results based on our three conditions (the three levels in the variable Condition)—see (2).

Spend some time examining the patterns in Fig. 4.1. Among other things, we can see that when a pause is added after NP_2 (i.e., the Low condition), participants' preference for low attachment goes up relative to the NoBreak condition. Conversely, if you look at cases where a pause is added after NP_1 (i.e., the High condition), participants' preference for low attachment goes down (again relative to the NoBreak condition). It's easier to see what's going on once we define a "baseline", and the NoBreak condition plays that role. Now we have some idea of the effect of Condition on participants' responses.

How about Proficiency? Unsurprisingly, our hypothetical learners (who are Spanish speakers) have a weaker preference for low attachment overall when compared to native speakers of English. At the same time, we can see that advanced learners are more similar to native speakers, insofar as their preference for low attachment is stronger than that of intermediate speakers (for the High and Low conditions).

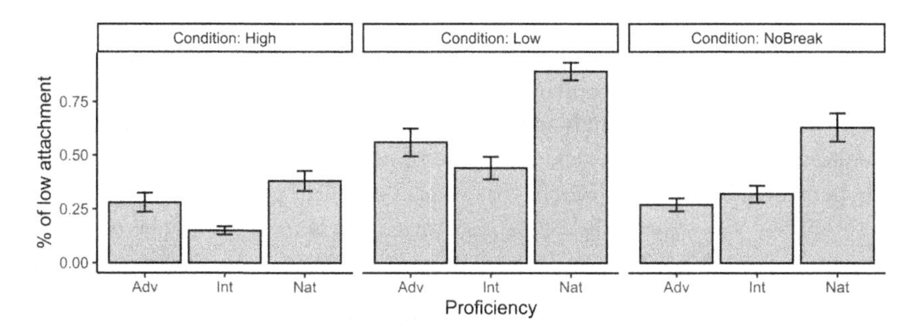

FIGURE 4.1 Resulting Plot after Data Transformation

The error bars in Fig. 4.1 give us some idea of what to expect when we analyze this data later (in chapter 7): error bars that don't overlap (or that overlap only a little) may reveal a statistical effect. For example, look at the bar for Adv in condition High, and compare that bar to Adv in condition NoBreak. These two bars overlap a lot (they have almost identical values along the *y*-axis). This suggests that advanced learners' preferences are not different between the High and NoBreak conditions. Here's another example: if we had to guess and had no access to any statistical method, we could say that "it's more likely that native speakers are different from advanced learners in the Low condition, than that advanced learners are different from intermediate learners in the same condition". The take-home message here is that Fig. 4.1 already tells us a lot about the patterns in our data. It helps us understand what's going on, and it also helps us move forward with our analysis.

Let's now examine the code block that prepared our data and generated Fig. 4.1. Recall that we're currently working with the categoricalDataPlots.R script. Code block 20 therefore mirrors what the top of that script should look like. Lines 2–7 are familiar at this point, as they simply load tidyverse and our data (rClauseData.csv) and visualize the top rows of the data—if you haven't closed RStudio since chapter 3, tidyverse is already loaded, but you can run line 2 anyway. The code is divided into two parts: first, we prepare the data (lines 10–17), and then we plot it (lines 25–30). Once you prepare the data, the plotting itself should be familiar.

Lines 10–17 create a new variable, which will be our new tibble with percentages. First, line 10 takes our data as the starting point (rc). Then, line 11 filters the data by removing all the fillers and focusing only on target items (recall that the column Type has two values, Filler and Target). Third, line 12 groups the data by a number of variables, namely, ID, Proficiency, Condition, and Response—these variables will be present in our new tibble, called props. Imagine that R is simply "separating" your data based on some variables, such that every participant will have his/her own dataset, and within each dataset R will further separate the data based on the variables we want. Fourth, line 13 counts the responses, that is, how many High and Low responses there are for each participant and condition.

At this point we have raw numbers only (counts), not percentages. In other words, props right now contains a column n where we have the number of times participants chose High and Low for each condition—and we also have a column for proficiency levels, as discussed earlier. Our next step is to calculate the percentages. The issue here is that we can calculate percentages in different ways. What we want is the percentage that represents the number of times someone chose High or Low *for a given condition*. That's what line 14 does: it groups the data again, this time by Proficiency, ID, and Condition. Why don't we have Response here? Assume a participant had five responses for High and five responses for Low. We want to divide each count by *both*

R code

```
# Remember to add this code block to categoricalDataPlots.R
library(tidyverse)

# Import data:
rc = read_csv("rClauseData.csv")

head(rc)

# Prepare data: percentages
props = rc %>%
  filter(Type == "Target") %>%
  group_by(Proficiency, ID, Condition, Response) %>%
  count() %>%
  group_by(Proficiency, ID, Condition) %>%
  mutate(Prop = n / sum(n)) %>%
  filter(Response == "Low") %>%
  ungroup()

# Visualize result (starting with ID s1):
props %>%
  arrange(ID) %>%
  head()

# Figure (bar plots + error bars):
ggplot(data = props, aes(x = Proficiency, y = Prop)) +
  stat_summary(geom = "errorbar", width = 0.2) +
  stat_summary(geom = "bar", alpha = 0.3, color = "black") +
  facet_grid(~Condition, labeller = "label_both") +
  labs(y = "% of low attachment") +
  theme_classic()

# ggsave(file = "figures/mainBinaryPlot.jpg", width = 7, height = 2.5, dpi = 1000)
```

CODE BLOCK 20 Preparing Binary Data for Bar Plot with Error Bars

responses; otherwise we'd be dividing a number by itself, the result of which would always be 1. That's why we don't group by Response in line 14. We simply take n, which was already there as a result of lines 10–13, and divide the counts *across* both responses. Because we're still grouping the data by Proficiency, ID, and Condition, this will give us the percentage of High and Low responses by participant and by condition.[4]

Line 15 in code block 20 now allows us to create a new column, which will divide the existing column n (calculated in line 13) by the sum of n across the variables in line 14. Finally, we remove High responses to keep only Low responses (line 16) and ungroup the data in line 17.[5]

Fig. 4.2 visually demonstrates the result of the process described earlier using one participant (ID s1) as an example. Being a native speaker, participant s1 is nested inside Nat. This participant chooses Low attachment 50% of the time when the condition is NoBreak. And because there are only two possible responses, we know that this participant also chooses High attachment 50% of the time for the same condition.

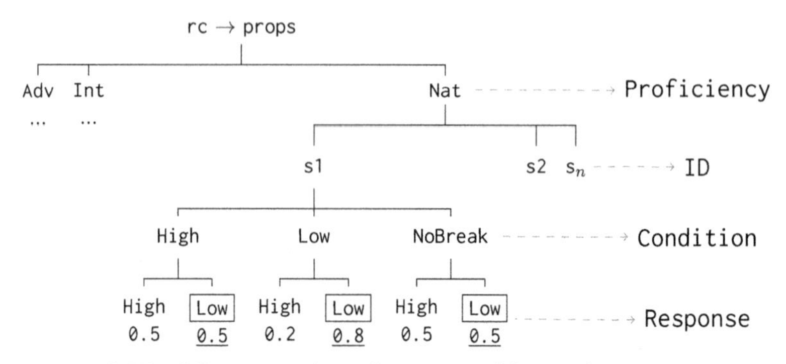

FIGURE 4.2 A Visual Representation of props: participant s1

Our new tibble, props, contains three rows per participant (the proportions for all three Low responses we care about). If you run lines 20–22 in code block 20, you will see that our participant (s1) in Fig. 4.2 fills the top three rows of the output (line 21 is sorting the data by ID). In column Prop, the values for s1 should be 0.5, 0.8, and 0.5—which are the values underlined in Fig. 4.2. Finally, because each participant now has three rows in our new data, and because we have 30 participants, props has 90 rows, right? Well, not exactly. You'll notice that props has 88 rows. Why is that? When we process our data the way we did, NAs are not counted. As a result, if a given participant only had High responses for a particular condition, that means our participant will have fewer than three rows in props—this is the case for participants s23 and s25, who have no Low responses for the High condition. That's why we ended up with 88 instead of 90 rows. There is a way to include empty counts, but we are not going to worry about that since it won't affect our analysis.

Once we prepare our data, creating Fig. 4.1 is straightforward, given what we discussed in chapter 3. After all, we now have a continuous variable, Prop, in props. Lines 25–30 are all we need to produce Fig. 4.1.

The nice thing of having ID as one of our grouping variables earlier is that we can now calculate not only the mean proportion of Low responses in the data but also the standard error across participants. You can go back to code block 20 and remove ID from lines 12 and 14. If you check the number of rows now, instead of 88 rows we have only 9—we have three proficiency levels and three conditions, and we're only looking at Low responses. The limitation here is that we wouldn't know how individual participants behave: we'd merely have the overall proportion of Low responses for *all* advanced learners for each condition, for example. That's why we should include ID as a grouping variable.

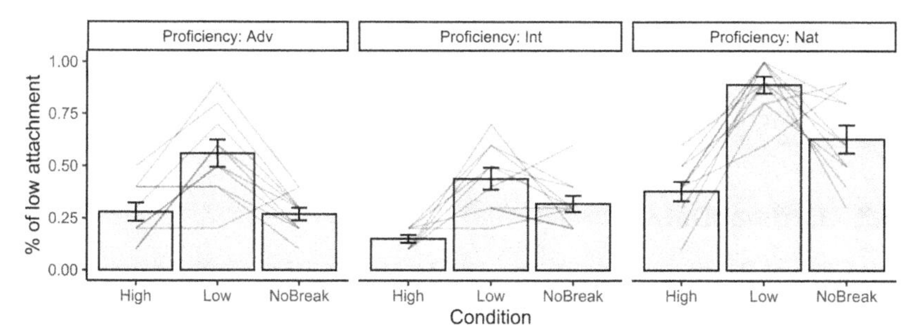

FIGURE 4.3 By-Speaker Responses

R code

```
# Figure (bar plots + error bars + by-participant lines):
ggplot(data = props, aes(x = Condition, y = Prop)) +
  stat_summary(geom = "line", alpha = 0.2, aes(group = ID)) +
  stat_summary(geom = "errorbar", width = 0.2) +
  stat_summary(geom = "bar", alpha = 0.1, color = "black") +
  facet_grid(~Proficiency, labeller = "label_both") +
  labs(y = "% of low attachment") +
  theme_classic()

# ggsave(file = "figures/varBinaryPlot.jpg", width = 7, height = 2.5, dpi = 1000)
```

CODE BLOCK 21 Adding by-Participant Lines to a Bar Plot

Fig. 4.3 is very similar to Fig. 4.1. Like before, we have bars representing means and error bars representing standard errors from the means. But now we also have lines that represent the average response by participant, so we can see how the variation across participants generates the error bars that we saw before in Fig. 4.1.[6] For example, compare condition High between Adv and Int and you will see why the error bars have different heights: in the High condition, advanced learners' responses are more similar than intermediate learners. Finally, notice that Fig. 4.3 has Condition on its *x*-axis and Proficiency across facets, so it's structurally different from Fig. 4.1. Stop for a minute and consider why our *x*-axis can't represent Proficiency in Fig. 4.3. The modification is necessary because a single participant sees all conditions in the experiment, but a single participant can't possibly have more than one proficiency level at once. Therefore, if we had Proficiency along our *x*-axis, we wouldn't be able to draw lines representing individual participants.

The code used to generate Fig. 4.3 is shown in code block 21. The key here is to examine line 3, which uses stat_summary()—but this time specifies a different geom. Transparency (alpha) is important to keep things clear for the reader—both lines and bars have low alpha values in the figure. Crucially, we specify aes(group = ID) inside our stat_summary(), which means each line will represent a different ID in the dataset (props). If you use color = ID in addition to group = ID, each participant will have a different color (and a key will be added to the right of our figure)—clearly this is not ideal, given the number of participants. Finally, note that if we hadn't added ID as a grouping variable in code block 20, we wouldn't have that column now, and Fig. 4.3 wouldn't be possible.

4.2 Ordinal Data

We often deal with ordinal data in second language research. We may ask participants to judge the naturalness of a sentence using a 4-point scale, for example. Our dataset rc actually has one ordinal variable: Certainty. Participants in our hypothetical relative clause study were asked to rate their level of certainty using a 6-point scale. In such a scale, only the end-points are labeled (1 = NOT CERTAIN, 6 = CERTAIN).

Let's assume for a second that all points along our scale were labeled. How could we label six degrees of certainty? Here's a possibility: NOT CERTAIN AT ALL, NOT REALLY CERTAIN, KIND OF CERTAIN, PRETTY CERTAIN, VERY CERTAIN, ABSOLUTELY CERTAIN. Here we run into two problems. First, it's harder than we think to come up with appropriate labels in this case. Perhaps we could reduce the number of levels to four instead of six (well, now it'd be too late for that). But even if we had four levels, we'd still run into a second more serious problem: points on a scale are equidistant. The distance from point 1 to point 2 must be the same as the distance from point 4 to point 5. Now look back at our attempt to label all points: can we say that the distance between NOT CERTAIN AT ALL and NOT REALLY CERTAIN is the same as the distance between PRETTY CERTAIN and VERY CERTAIN? How can we even answer that question?

Alright, enough about points on a scale—we will explore them in more detail in chapter 8. Here, we want to plot Certainty in our rc tibble. You may be wondering whether we could simply assume that this variable is continuous—after all, we have a range going from 1 to 6. For plotting the data, we *could* assume that. You could, for example, create a simple bar plot with error bars where y = Certainty and x = Proficiency. That will work, and you should definitely try it.

As you create your bar plot, you'll use stat_summary() to generate both the bars and the error bars. A potential issue here, however, is that scalar or ordinal data points are *rarely* normally distributed—you can check that by creating a histogram, which will show a uniform distribution in this case (refer to Fig. 3.2). Sometimes your ordinal data will look bimodal (many responses for 1 and many responses for 6, but little in between). Sometimes your distribution will look skewed (e.g., participants were all quite certain, it seems, so no one chose points below 4, and now they look clustered around 5 and 6). The bottom line is: if you have a scale, your distribution will almost never be normal. As a result, looking at means and standard deviations (or standard errors) can be a little tricky.

When we create a bar plot with error bars for an ordinal variable, we are accepting that using means and standard errors may not be the best way to examine our data. One way to be more conservative would be to bootstrap our standard errors (§1.3.4). stat_summary() assumes by default that you want geom = "pointrange" and that your variable is normally distributed. If your variable is not normally distributed, you can use stat_summary(fun.data = mean_cl_boot). This function bootstraps error bars (see §1.3.4)—you will notice, if you compare both methods, that this increases the size of your error bars.[7] The key is to assess how reliable the figure is given the statistical patterns in the data, which we will examine in chapter 8. If the appropriate statistical model is consistent with the patterns shown in a bar plot (even in one with traditional standard errors), then our figure is fine.

Our current hypothetical study is clearly not centered around Certainty—our main variables of interest were introduced and visualized in §4.1. As a result, certainty levels here are merely extra pieces of information regarding our participants' behavior, and for that bar plots with standard errors are enough. What else could we do with ordinal data, though?

There's another common plot that is more aligned with the ordinal nature of our variable Certainty. It's also a bar plot, but it's a different kind of bar plot. Fig. 4.4 plots certainty levels with stacked bars. Different shades of gray represent different points on our 6-point scale. The height of each bar tells us how representative each certainty level is for each proficiency level and for each condition. This type of figure allows us to see where participants' Certainty responses are clustered. For example, native speakers seem more certain in the Low condition than in the other two conditions. Learners, in contrast, seem more certain in the High condition than in the other two conditions—we can easily see that by looking for the highest concentration of dark bars in the plot, because darker shades of gray here represent a higher level of certainty. This makes our figure intuitive, and we don't have to worry about the non-normality of our variable. We do, however, have to compute the percentages being plotted, which resembles what we did for our binary data in Fig. 4.1.

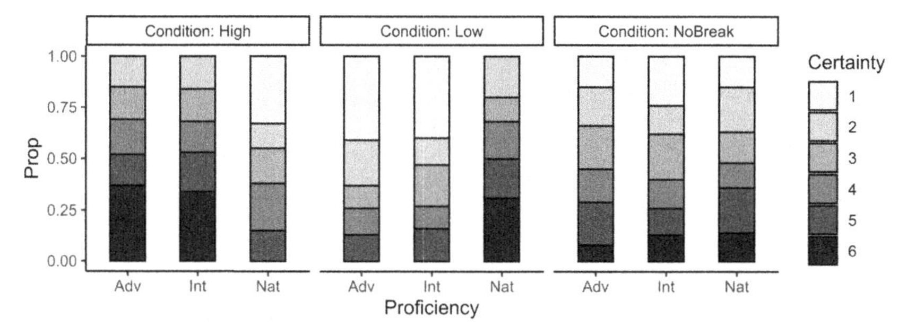

FIGURE 4.4 Plotting Ordinal Data with Bars

Code block 22 provides all the code you need to generate Fig. 4.4. The first thing we need to do is check whether R understands that our Certainty variable is, in fact, an ordered factor (i.e., not just a continuous variable). Recall that by running str(rc) we have access to the overall structure of our data, including the class of each variable. If you run line 2, you will see that R treats Certainty as a variable containing integers (int). This is expected: R can't know whether your numbers are true numbers or whether they come from a scale with discrete points. Lines 5–6 fix that by recategorizing the variable: now Certainty is an ordered factor.

Lines 9–14 prepare our data by computing counts and proportions for each point on our scale. Most of it will be familiar from code block 20, which we discussed at length in §4.1. This time we don't include ID as one of our variables of interest because we're not planning to plot by-participant trends (and we have no error bars in Fig. 4.4)—in chapter 5, we will revisit Fig. 4.4 to improve different aspects of our data presentation.

Bar plots are extremely useful to display ordinal variables—they are certainty not the only way to plot such variables, but they do communicate our results in a clear and familiar way. As we saw earlier, you can treat your scale as a continuous variable, in which case traditional bar plots and standard errors may be used (with care) to plot the mean certainty in your data. Alternatively, you can compute percentages for responses along your scale and use those percentages to create a stacked bar plot where fill colors can help us focus on clusters of

R code

```
1  # Check if Certainty is an ordered factor:
2  str(rc)
3
4  # Make Certainty an ordered factor:
5  rc = rc %>%
6    mutate(Certainty = as.ordered(Certainty))
7
8  # Prepare our data (generate percentages):
9  cert = rc %>%
10   filter(Type == "Target") %>%
11   group_by(Proficiency, Condition, Certainty) %>%
12   count() %>%
13   group_by(Proficiency, Condition) %>%
14   mutate(Prop = n / sum(n))
15
16 # Create bar plot:
17 ggplot(data = cert,
18        aes(x = Proficiency,
19            y = Prop,
20            fill = as.ordered(Certainty))) +
21   geom_bar(stat = "identity", width = 0.5, color = "black") +
22   facet_grid(~Condition, labeller = "label_both") +
23   theme_classic() +
24   scale_fill_brewer("Certainty", palette = "Greys", direction = 1)
25
26 # ggsave(file = "figures/certaintyPlot.jpg", width = 7, height = 2.5, dpi = 1000)
```

CODE BLOCK 22 Preparing the Data and Plotting Certainty Levels

responses along the scale (as opposed to using means). If your scalar data plays a key role in your study, the latter method is recommended as you don't have to worry about the distribution of your ordinal data—which is almost never normal.

As we saw earlier, visualizing ordinal data is not so different from visualizing binary data. Both tend to require some data preparation before running ggplot ()—in that sense, both data types can be more demanding than continuous data. Fortunately, you can use the code examined earlier as a template for future reference—adjusting some code is almost always easier than creating the same code from scratch.

4.3 Summary

In this chapter, we discussed how to prepare our categorical data for plotting. As we saw, when dealing with binary data we will often transform our data into counts or percentages, which in turn allows us to employ the same visualization techniques discussed in chapter 3. Here's a brief summary of the chapter.

- To group our data by different variables, we can use group_by(). Every time we want to analyze a variable taking into account different groups (or variables), we will need to first group our data. That's why group_by() is such an important function in the code blocks discussed in this chapter. Fig. 4.2 illustrates how grouping works across different variables.
- For binary response variables such as High and Low in the hypothetical relative clause study explored earlier, we will often delete one of the two options at the end of our data preparation. In other words, once we calculate the percentages for each response, we will likely plot only one response (e.g., % of low responses, as shown in Fig. 4.1).
- When we have a variable in our data with scalar data, for example, a range from 1 to 6, R will categorize said variable as continuous (either numeric or integer).
- For ordered factors (ordinal or scalar data), we examined two options: first, we can treat such variables as continuous and use bar plots accompanied by standard error bars—but remember that scalar data is rarely normal. In that case, we don't need to change the class of our variable (see previous point). Alternatively, we can use bar plots and fill colors to plot percentages of responses along the scale. In that scenario, shown in Fig. 4.4, we have to first transform our scalar variable into an ordered factor using as.ordered()—refer to code block 22, lines 5–6.

4.4 Exercises

Ex. 4.1 Changing Fig. 4.1

1. Modify the code used to generate Fig. 4.1 so that the *y*-axis shows the percentage of high attachment.

R code
```
1  # Plotting by-speaker results:
2
3  ggplot(data = props %>% filter(Proficiency %in% "Nat", "Int"),
4         aes(x = reorder(ID, Prop), y = Prop, color = proficiency)) +
5    stat_summary()
6    labs(x = "Speaker", y = "Proportion of Low responses") +
7    theme_classic()
```

PROBLEMATIC CODE C

2. Generate Fig. 4.1 without native speakers (by filtering the data in the plot itself). *Hint:* you can use the inequality operator (!=).
3. Create a box plot version of Fig. 4.1 and add means and standard errors to it—you should use props, created in code block 20. *Hint:* if you simply add stat_summary() to your figure, it will plot means and standard errors (in the form of point ranges).

Ex. 4.2 Plotting by-Item Variation

1. It's sometimes useful to visualize by-participant patterns—even though we're mostly interested in generalizing group-level patterns in our data. Problematic code C attempts to create a plot where non-native participants are on the x-axis and the proportion of Low responses is plotted along the y-axis. Your task is to fix the code and successfully create the figure.

Notes

1. Unless you have coded yes as 1 and no as 0, in which case the proportion of yes responses is easy to calculate.
2. Fillers in this study could be sentences that are not semantically ambiguous or sentences where the focus is not the head of a relative clause and so on.
3. R orders level labels alphabetically. We will see how to change that in chapter 5.
4. We're also grouping by Proficiency, but because a participant cannot have two different proficiency levels, this will not affect how we calculate the percentages.
5. By default, only non-zero counts are included in the resulting tibble. If your dataset has zero counts for certain combinations, not including them may impact calculations based on the Prop column (e.g., means), which may in turn affect any figures that depend on these calculations. To avoid this problem, you should use the complete (x, y) function, where x represents the response variable of interest and y represents a list where we specify how we want to fill in zero counts and proportions. For example, y could be fill = list(n = 0, Prop = 0) in this case.
6. This is an example where we are using lines even though our x-axis is a discrete variable (cf. §3.7). Here, however, lines play a different role, connecting the means of each participant across conditions—otherwise it would be difficult to visually trace the trends in question.
7. If the error bars don't change much, you could keep the traditional error bars.

5

AESTHETICS: OPTIMIZING YOUR FIGURES

Thus far we have examined several plots created with ggplot2. We have focused our discussion on the *content* of each figure, so we ignored aesthetic characteristics such as font size, label orders, and so on. These characteristics are the focus of this chapter. Fortunately, ggplot2 gives us total control over the aesthetics of our plots, which is important to generate publication-quality figures.

Before we get started, go ahead and create a new script, optimizingPlots.R, which will be our third script inside the directory plots in Fig. 3.1. Our ultimate goal in this chapter is to create Fig. 5.1. In doing that, we will explore a number of aesthetic modifications that can be made using ggplot2. Plotting ordinal data is a little trickier than plotting continuous data, so this figure will be good practice for us.

You can compare Fig. 5.1 with our first attempt to plot certainty levels back in Fig. 4.4. Here, we have important differences. First, our bars are now horizontal, not vertical, so they mirror the horizontal scale in the experiment. Second, we no longer have a key to the right of the figure. Instead, we moved the certainty points onto the bars themselves. As a result, you don't have to keep looking back and forth to remember which level each color represents. Note, too, that the font color of the scale points alternates between black and white in order to maximize the contrast between the certainty values and the fill of the bars. Third, we finally have actual percentages on our *y*-axis—only now we have flipped our axes, so our *y*-axis is our *x*-axis and vice versa. Fourth, we have changed the labels for Proficiency (*y*-axis), which are no longer abbreviated. We have also rotated those labels, so they are parallel with the plot's border and thus occupy less space on the page. The labels for Condition also have a minor change: condition NoBreak is

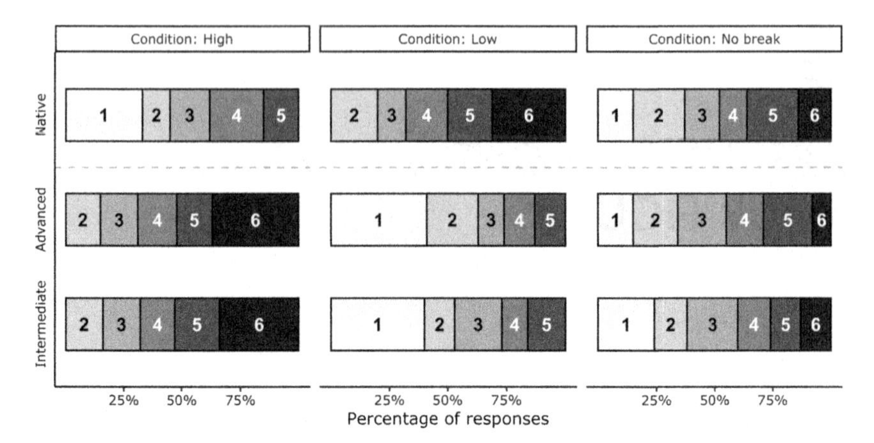

FIGURE 5.1 Plotting Ordinal Data with Adjustments

now No break, with a space. Finally, we now have a horizontal dashed line separating native speakers from learners—and, you may not have noticed this, but we now have different font family (Verdana) and size. As you can see, a lot is different here. We will spend the remainder of this chapter examining in detail how to implement all these changes in R. Most of the changes we will discuss are applicable to all the other plots we have seen so far, as long as the data and plot you have in mind are compatible with the specification you add to your code.

We start with some data preparation, shown in code block 23. Because this is a new script, we first need to load some packages. This time, we'll need more than just tidyverse, which means you will have to install two more packages: scales and extrafont. You can install both packages by running install.packages(c("scales", "extrafont")). The first package, scales, allows us to adjust our axes based on different scales (it will give us the percentages along the axis). The second package, extrafont, will allow us to change the font family in our figures—most of the time this isn't necessary, since the default font family is more than appropriate, but some journals will require a specific font for your figures.

Once you have installed our two new packages, you're ready to run lines 1–4 in code block 23—line 4 simply loads all the fonts available on your computer by using extrafont, so a list of fonts will be printed in your console. You can check which fonts you can use by running line 7. Lines 9–14 are familiar at this point: we're importing our data and making the variable Certainty an ordered factor.

Lines 17–22 compute proportions. Simply put, we want to know what percentage of responses we have for all six points along our certainty scale. Crucially, we want the percentages to be computed based on participants' proficiency levels and the three experimental conditions (High, Low,

R code
```
library(tidyverse)
library(scales)    # To get percentages on axis
library(extrafont)
loadfonts()

# Check which fonts are available:
fonts()

# Import data:
rc = read_csv("rClauseData.csv")

# Make Certainty an ordered factor:
rc = rc %>%
  mutate(Certainty = as.ordered(Certainty))

# Prepare our data (generate percentages):
cert = rc %>%
  filter(Type == "Target") %>%
  group_by(Proficiency, Condition, Certainty) %>%
  count() %>%
  group_by(Proficiency, Condition) %>%
  mutate(Prop = n / sum(n))

cert = cert %>%
  ungroup() %>%
  mutate(color = ifelse(Certainty %in% c(4, 5, 6), "white", "black"),
         Proficiency = factor(Proficiency,
                              levels = c("Int", "Adv", "Nat"),
                              labels = c("Intermediate", "Advanced", "Native")),
         Condition = factor(Condition,
                            levels = c("High", "Low", "NoBreak"),
                            labels = c("High", "Low", "No break")))
```

CODE BLOCK 23 Plotting Certainty Levels with Adjustments: Preparing the Data

NoBreak), that's why we're using group_by() once again—we do that in lines 19–20 to generate a count and again in 21–22 to compute proportions (this process should be familiar by now). The result of lines 17–22 is that we now have a new variable, cert, which contains five columns: Proficiency, Condition, Certainty, n, and Prop.

Lines 24–32 are the chunk of code that is actually new in code block 23. Here we're making some adjustments to cert. First, we ungroup the variables (which we grouped twice in lines 17–22). We then create one column (color) by using a conditional statement. The column color will have the following values: "white" if a given certainty level is 4, 5, or 6 and "black" otherwise. We then modify the labels of two existing columns (Proficiency and Condition). In lines 27–29, we are asking R to take the existing levels of Proficiency and label them with non-abbreviated proficiency levels. In lines 30–32, we are simply adding a space to NoBreak in Condition. This is all we need to do to get started with our figure.

Code block 24 is longer than usual, mostly because of our several adjustments. The layers here are (mostly) ordered from major (content-related) to minor (form-related). You can probably guess what every line of code in code block 24 is doing. That's the advantage of having access to the actual code. You don't ever have to start the figure from scratch: you can adapt the code provided here to suit your data.

Our first layer (lines 2–6) for Fig. 5.1 points ggplot2 to our data object (cert), as usual, but this time we have four arguments inside aes(). In addition to our typical axes, we also establish label = Certainty, and fill = Certainty. The former is responsible for labeling the points along the scale; the latter changes the fill color of the bars depending on the certainty value (this should be familiar from Fig. 4.4). Next, in lines 7–8, we add our bars. We specify stat = "identity" because we are providing the y-axis ourselves, and we want the bars to plot those values (and not, say, calculate the values some other way). There we also adjust the width of the bars and the color for their borders. Line 8 is important: we're asking ggplot2 to reverse the order of the scale—this will ultimately guarantee that our bars go from left (1) to right (6), and not the other way around, which would be counterintuitive.

Next, we flip the entire figure in line 9 by using coord_flip()—this is what makes our y-axis become our x-axis and vice versa. Naturally, you can flip other plots in ggplot2 by adding this layer to your code.

R code

```
1  # Create bar plot for ordinal data:
2  ggplot(data = cert,
3         aes(x = Proficiency,
4             y = Prop,
5             label = Certainty,
6             fill = Certainty)) +
7    geom_bar(stat = "identity", width = 0.5, color = "black",
8             position = position_stack(reverse = TRUE)) +
9    coord_flip() +
10   geom_text(aes(color = color),
11             position = position_fill(vjust = 0.5, reverse = TRUE),
12             fontface = "bold") +
13   facet_grid(~Condition, labeller = "label_both") +
14   geom_vline(xintercept = 2.5, linetype = "dashed", color = "gray") +
15   scale_fill_brewer("Certainty", palette = "Greys", direction = 1) +
16   scale_color_manual(values = c("black", "white")) +
17   scale_y_continuous(labels = percent_format(), breaks = c(0.25, 0.5, 0.75)) +
18   theme_classic() +
19   theme(legend.position = "none",
20         text = element_text(family = "Verdana", size = 10),
21         axis.ticks.y = element_blank(),
22         axis.text = element_text(color = "black"),
23         axis.text.y = element_text(angle = 90, hjust = 0.5)) +
24   labs(y = "Percentage of responses",
25        x = NULL)
26
27 # ggsave(file = "figures/certaintyPlot2.jpg", width = 7, height = 3.5, dpi = 1000)
```

CODE BLOCK 24 Plotting Certainty Levels with Adjustments: Creating Plot

Once our axes are flipped, lines 10–12 adjust the text in our labels, which represent the certainty levels in Fig. 5.1. First, we add aes(color = color), which specifies that the color of the text depends on the value of the column color in our data. Line 11 then adjusts the vertical position of the labels[1]—try removing vjust from line 11 to see what happens. Line 12 defines that the labels along the scale should be bold.

Line 13 is familiar: we're adding another dimension to our figure, namely, Condition—and we're asking ggplot2 to label the values of the variable. Next, line 14 adds the horizontal dashed line separating native speakers from learners. The function, geom_vline() draws a *vertical* line across the plot, but because we have flipped the axes, it will draw a horizontal line instead. We still have to give it an xintercept value (i.e., where we want to the line to cross the *x*-axis), even though the axes have been flipped. Because our axis contains three discrete values, xintercept = 2.5 will draw a line between levels 2 and 3, that is, between advanced learners and native speakers.

Lines 15–17 use similar functions. Line 15, scale_fill_brewer() is using a predefined palette (shades of gray) to fill the bars (recall that the fill color will depend on the variable Certainty, which we defined back in line 6).[2] The argument direction merely defines that we want the colors to follow the order of the factor values (direction = −1 would flip the order of colors). As a result, darker colors will represent a higher degree of certainty. Line 16 *manually* defines the colors of the labels, and line 17 defines not only that we want percentages as labels for our former *y*-axis (now *x*-axis) but also that we want to have only three breaks (i.e., we're omitting 0% and 100%, mostly because the *x*-axis looks cluttered across facets if we don't omit those percentage points).

Everything from line 19 to line 23 is about formatting. First, we get rid of our key (the one we have in Fig. 4.4), since now we have moved the actual certainty values onto the bars for better clarity. Second, in line 20, we choose Verdana as our font family, and we set the font size we desire.[3] Line 21 removes the ticks along our *y*-axis, line 22 changes the color of our axis text (it's typically dark gray, not black), and line 23 changes the angle of the labels along our *y*-axis.

In theory, we could remove the percentage points from the *x*-axis in Fig. 5.1. The way the figure is designed, we just have to look at the sizes and colors of the bars, and we will easily see where participants are more or less certain. The actual percentage point of each bar is clearly secondary, and by definition, all bars must add up to 100%. Finally, note that the *y*-axis is not labelled. The idea here is that its meaning is obvious, so no labels are necessary (you may disagree with me).

You might be wondering about the title of the figure. Even though we can include a title argument in labs(), I personally never do that. Titles are usually defined not in the plot itself, but rather in the caption of the plot, which is defined later (e.g., in the actual paper)—this is the case for all figures in this book.

5.1 More on Aesthetics

You may have seen figures in published papers where *p*-values and comparisons are shown on top of bar plots or box plots, for example. They will often show horizontal square brackets across different categories being compared, so the figure shows the data *and* the statistical results (to some extent). You can easily do that in R by installing the ggpubr package and then adding the following line to your figures: stat_compare_means()—read about the package at https://cran.r-project.org/web/packages/ggpubr/index.html. If you wish to add *p*-values and comparisons to your plot, however, bear in mind that (i) this will likely focus your results on *p*-values (as opposed to effect sizes), and (ii) the more comparisons you have, the more cluttered your plot will look. Indeed, showing comparisons and *p*-values in a figure is likely a tradition inherited from old statistics, where the focus of the analysis was on *p*-values and on rejecting the null hypothesis.

(i) and (ii) are the reasons that I do not show *p*-values for comparisons in figures in this book. Instead, figures to explore the data and statistical results will be *separate* components of the analyses in the chapters to come. You will definitely see figures showing statistical results, but those will be separate from exploratory figures—and, crucially, will focus on effect sizes, not on *p*-values.

In summary, ggplot2 offers a lot of flexibility: we can create virtually any static figure we want, given that we have total control over the aesthetic parameters involved in a typical figure—see Wickham (2016) for a comprehensive guide or visit https://ggplot2-book.org/index.html. If you wish to go beyond the typical aesthetic features offered by ggplot2, you may want to explore packages such as ggthemes and ggtext. Fortunately, once you create the code for your ideal figure, you can easily reproduce said figure multiple times by simply adjusting the code—as discussed in *How to use this book* back in the preface of this book. Ultimately, much like R in general, you can always adjust an existing ggplot2 code instead of creating a figure from scratch—which is especially helpful when you're learning how to use R.

5.2 Exercises

Ex. 5.1 Colors and Fonts

1. Install and load the RColorBrewer package. Then run display.brewer.all() on your console. A collection of preset palettes will appear in your plot window (pane D). Play around with different palettes to add some color to Fig. 5.1.
2. Run fonts() in your console to see which fonts you can choose from. Try recreating Fig. 5.1 with different font families and sizes to get used to the

theme() layer. Run ?theme or help(theme) in your console to learn more about it.

3. Add the following layer to Fig. 5.1: scale_x discrete(label = abbreviate). This will abbreviate labels along the x-axis, which can be quite useful at times. You can always manually specify labels, of course.

Notes

1. vjust adjusts the vertical justification of the bars (options: 0, 0.5, 1)—0.5 centers the labels.
2. ggplot2 has its own default colors, but you will likely want to change them. There are numerous palettes available when you use scale_fill_brewer(). One example is the palette RdOr, which goes from red to orange and can be quite useful for plotting scalar data. In addition to Greys, popular palettes include Blues and Greens.
3. Technically, you can change the font size simply by changing the width and height of the plot when you save it in line 26. For example, if you set weight to 5 and height to 2, your font size will be much larger—too large, in fact.

PART III

Analyzing the Data

6

LINEAR REGRESSION

In Part II we visualized different datasets and explored different figures that communicate the patterns in our data. The next step is to focus on the statistical analysis of our data. Even though we discussed trends in the data as we created different figures, we still don't know whether such trends are statistically relevant.

In this chapter, we will focus on linear regressions. To do that, we will return to continuous (response) variables, which we visually examined in chapter 3. Linear regressions will be the first of the three statistical models we discuss in this book. Later, in chapters 7 and 8, we will explore categorical response variables, which we visually examined in chapter 4. As you can see, we will follow the same structure used in Part II, where we started our discussion with continuous data and later examined categorical data.

For the present chapter, we will start with an introduction to linear regressions (§6.1). Then, in §6.2, we will statistically analyze the data discussed in chapter 3. Finally, we will spend some time on more advanced topics (§6.3).

Given that our focus in this book is on regression analysis, this chapter will be the foundation for chapters 7 and 8. Before we get started, however, it's time to discuss file organization one more time. So let's make sure all our files are organized in a logical way.

In Fig. 3.1 (chapter 3), the folder bookFiles contains two subfolders, basics and plots. Recall that the plots folder contains all the files we used in Part II of this book. We will now add a third subfolder to bookFiles—let's call it models. This folder will in turn contain another subfolder, Frequentist. Inside Frequentist, we will add all the files for chapters 6–9, and we'll also create another figures folder, just like we did in the basics and plots directories. Later, when we discuss Bayesian models (chapter 10), we will add another subfolder to models, namely, Bayesian—remember that you can check the file structure used in this book in Appendix D.

For this chapter, we will first create an R Project called Frequentist.RProj (inside the Frequentist folder/directory), and then we will add three new scripts to our Frequentist directory—simply create three new empty scripts once you've created the R Project and save them as dataPrepLinearModels.R, plotsLinearModels.R, and linearModels.R. To recap, you should have the following folder structure: bookFiles > models > Frequentist. Inside Frequentist, you should have one folder (figures) and four files: an R Project (Frequentist.RProj) and three scripts (dataPrepLinearModels.R, plotsLinearModels.R, and linearModels.R). At this point you can probably guess where this is all going: we will prepare our data in dataPrepLinearModels.R, plot some variables in plotsLinearModels.R, and work on our statistical analysis in linearModels.R. As mentioned earlier, here we will use the same dataset from chapter 3, namely, feedbackData.csv. Go ahead and add a copy of that file to the newly created folder Frequentist.

Recall that our feedbackData.csv is a wide table, so we need to transform the data before plotting or analyzing it (the same process we did for figures in Part II). Fortunately, we have already done that in code block 11. So all you need to do is copy that code block, paste it into dataPrepLinearModels.R, and then save the file. Next, open plotsLinearModels.R and add source("dataPrepLinearModels.R") to line 1. Once you run that line, you will load the script in question, which will in turn load the data file feedbackData.csv as feedback and create a long version of feedback, longFeedback. All the figures shown later should be created inside plotsLinearModels.R, *not* linearModels.R, which is reserved only for the actual statistical models. The code blocks that generate figures will save them in the figures folder as usual—feel free to change the location if you prefer to save the figures elsewhere on your computer.

Next, open linearModels.R and add source("dataPrepLinearModels.R") to line 1 as well, so both plotsLinearModels.R and linearModels.R will source the script that imports and prepares the data. All the code blocks that run models shown later should be added to linearModels.R. This should be familiar at this point.[1]

To make things easier, make sure you have all three scripts we just created open in RStudio (in three different tabs), so you can easily go from data preparation to plots to models. This is a very effective way to work with data: we separate these three essential components into three separate scripts in a self-contained project in the same directory. In this chapter, we will focus on linearModels.R, but it's nice to be able to generate our figures in plotsLinearModels.R—you can naturally copy some of the code blocks from chapter 3 into plotsLinearModels.R.

DATA FILE

In this chapter (§6.2) we will use feedbackData.csv again. Recall that this file simulates a hypothetical study on two different types of feedback, namely, explicit correction and recast. The dataset contains scores for a pre-, post-, and delayed post-test. Three language groups are examined, speakers of German, Italian, and Japanese.

6.1 Introduction

In this chapter, we will first review what linear models are and what they do (this section). We will then see them in action with different examples: you will learn how to run them in R, how to interpret them, and how to report their results. Once you finish reading the chapter, you may want to reread this introduction to review key concepts—a summary is provided at the end of the chapter, as usual.

Observe the scatter plot on the left in Fig. 6.1. There seems to be a relationship between the two variables in question: as we increase the value along our x-axis, we also see some increase in values on our y-axis. This pattern indicates some positive effect of x on y. On the right, we are fitting a line to the data. The goal here is simple: we want to use some explanatory variable (x-axis) to predict our response variable (y-axis). Simply put, we want to fit a line to the data that "best captures the pattern we observe". But *how* can we do that? And *why* would we want to do that? Let's focus on the *how* first.

What would a perfect line look like in Fig. 6.1? Well, it would be a line that touches *all* data points in the data. Clearly, a straight line is impossible here: our data points are too spread, so no straight line will be perfect. Here the best line is simply the line that reduces the overall distance between itself and all the data points. We can measure the dashed lines in Fig. 6.1 and use that as a proxy for

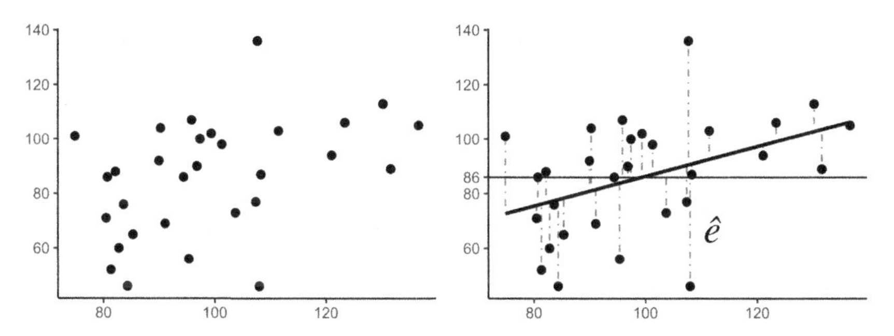

FIGURE 6.1 Fitting a Line to the Data

how much our line deviates from the actual data points. That distance, represented by $\hat{e}$ in the figure, is known as *residual*. Simply put, then, the residual is the difference between what our line predicts and the data observed—it's our *error*: $\hat{e} = \hat{y} - y$. We represent the true value of the error as ϵ and use $\hat{e}$ for the estimated value of the error.[2] $\hat{y}$ is the value we predict (i.e., our fitted line), and y is the actual data point in the figure.

If we sum all the residuals in Fig. 6.1, that is, all the vertical dashed lines, we will get zero ($\sum \hat{e} = 0$), since some distances will be negative and some will be positive (some data points are below the line and some are above it). If the sum of the residuals always equals zero, it's not a very informative metric, because *any* line will give us the same sum. For that reason, we square all the residuals. That way, each distance $\hat{e}^2$ will be a positive number. So now we can sum all these residuals squared ($\sum \hat{e}^2$) and that will give us a number representing how good our line is. If we try drawing ten different lines, the line that has the *lowest sum of squares of the residuals* is the best fit. The thick inclined line in Fig. 6.1 is the best line we can fit to our data here.

The horizontal solid line simply marks the mean of the variable on the y-axis ($\bar{y} = 86$) and serves as our baseline, since it ignores the variable on the x-axis. Our question, then, is how much better is the thick inclined line relative to our baseline?

One way to summarize a good fit is to use the *coefficient of determination*, or R^2. This number tells us what proportion of variance observed in the data is predictable from our variable on the x-axis of Fig. 6.1. R^2 ranges from 0 (no relationship between x and y) to 1 (perfect relationship/fit). In practice, perfect fits don't exist in linguistics, since patterns in language and speakers' behaviors are always affected by numerous factors—many of which we don't even consider in our studies. So the question is what constitutes a good fit, what is a good R^2. There's no objective answer to that question. For example, for the fit in Fig. 6.1, $R^2 = 0.15$. This means that the variable on the x-axis explains 15% more variation in the data relative to our flat line at the mean. You may think that 0.15 is a low number—and you're right. However, bear in mind that we are only considering *one* explanatory variable. If you can explain 15% of the variation with a single variable, that's definitely not a bad fit considering how intricate linguistic patterns are. The conclusion here is that x is useful in Fig. 6.1. In other words, using x is better than simply using the mean of y (our baseline)—among other things, this is what a statistical model will tell us.

You don't need to know how we calculate R^2, since R will do that for you. But the intuition is as follows (feel free to skip to the next paragraph if you already know how to calculate R^2): we compare the sum of squares for the flat line to the sum of squares for the fitted line. First, we assess our flat line at the mean by calculating how far it is from the actual data points:

$\sum(y-\bar{y})^2$, which is the same as $\sum\hat{e}^2$. This is our *total sum of squares*, or SS_t—the same process discussed earlier for the residuals, only now we're using a flat line at the mean as our reference. Next, we do the same process, this time comparing our fit line (in Fig. 6.1) to the actual data points (i.e., the sum of squares). This is our regression sum of squares (the residuals discussed earlier), or SS_r. We can then define R^2 as follows: $R^2 = \frac{SS_t - SS_r}{SS_t}$. Now think about it: if SS_r is zero, that means that we have a perfect line, since there's no deviation from the line and actual data points. As a result, $R^2 = \frac{SS_t - 0}{SS_t} = 1$. Conversely, if our line is just as good as drawing a flat line at the mean of y, then SS_t and SS_r will have the same value and will equal zero: $R^2 = \frac{SS_t - SS_r}{SS_t} = \frac{0}{SS_t} = 0$ (no relationship between x and y). Finally, you may be familiar with r, the *correlation coefficient*, which ranges from -1 to $+1$. R^2 is simply that number squared. Consequently, a correlation of ± 0.5 is equivalent to $R^2 = 0.25$.

Now that we know the intuition behind the notion of a "best" line, let's discuss why that's relevant. Once we have a line fit to our data, we can (i) examine whether x and y are correlated, (ii) determine whether that correlation is statistically significant, and (iii) predict the y value given a new x value (e.g., machine learning). For example, suppose we are given $x = 90$—a datum not present in our data in Fig. 6.1. Given the line we have, our y will be approximately 80 when $x = 90$. In other words, fitting a line allows us to predict what new data will look like.

Besides its predictive power, a linear model (represented here by our line) also allows us to estimate the relationship(s) between variables. For example, in our dataset from chapter 3, we could ask ourselves what the relationship is between the number of hours of study (`Hours` variable in `feedbackData.csv`) and participants' scores (`Score`). We will examine this particular example in §6.2.

When *can* we use linear models to analyze our data? If we have two continuous variables such as those shown in Fig. 6.1, the relationship between them must be linear—that's why you want to plot your data to inspect what kind of potential relationship your variables have. If your two variables have a nonlinear relationship, fitting a straight line to the data will naturally be of little use (and the results will be unreliable to say the least). For example, if your data points show a U-shaped trend, a linear model is clearly not the right way to analyze the data.

Two important assumptions on which a linear model relies are (i) that residuals should be normally distributed and (ii) that variance should be constant. Let's unpack both of them. Our residuals represent the *error* in our model—typically represented as ϵ. Some residuals will be positive, some will be negative, as discussed earlier. Some will be small (closer to our line), and some will be farther away. Indeed, if we created a histogram of the residuals, the histogram should look approximately normal: `ggplot(data = tibble(res = residuals(MODEL)), aes(x = R)) + geom_histogram()`.[3]

Besides creating a histogram to check whether the residuals are normally distributed, we could create a normal Q-Q plot, a special type of scatter plot that compares the quantiles of two distributions (a normal distribution against the distribution of residuals in this example). If the residuals are normally distributed, the data points in our Q-Q plot should follow a straight diagonal line. You can run such a plot in R as follows: `ggplot(data = tibble(R = residuals(MODEL)), aes(sample = R)) + stat_qq() + stat_qq_line()`—you can test this code with the models we will run in §6.2.

The second assumption, that variance should be constant, is also referred to as **homoscedasticity**. Look back at Fig. 6.1. Notice that the distances between the line and the data points (i.e., the residuals or errors) are relatively constant across the x-axis. Imagine that all the data points were clustered very close to the line at first, say between $x = 80$ and $x = 100$, and then spread apart for higher values of x. That scenario would violate the assumption that variance should be constant.

There are tests that check for the normality assumption (see Ghasemi and Zahediasl 2012 for a review). However, visually inspecting the residuals is likely the most practical way to make sure that our model is not violating its assumptions. Therefore, having a histogram of the residuals is already very effective to check for the normality assumption. Homoscedasticity is something we can often check during the data visualization stage of our analysis.

We now know that linear models are essentially lines fit to our data, and lines can be represented with *two* parameters: an intercept (β_0) and a slope (β). The intercept is where the line crosses zero on the x-axis—we can't see that in Fig. 6.1 because all the values are greater than or around 80. The slope tells us how inclined the line is. For example, if our line were flat (completely horizontal), its slope would be zero, which in turn would tell us that the explanatory variable on the x-axis doesn't affect our response variable on the y-axis. Simply put, the more inclined the line, the larger the effect size of the explanatory variable on the response variable.

As we saw earlier, everything that is *not* captured by our line is the error (ϵ) of the model represented by the line. The functions in 6.1 and 6.2 summarize our discussion thus far: $\hat{y}$ represents the predicted y value for participant i, our response variable on the y-axis in Fig. 6.1—we add hats to our terms to differentiate population parameters from sample parameters estimated by our model. We can't know the true values for β_0, β, and ϵ. But we can estimate them, $\hat{\beta}_0$, $\hat{\beta}$, and $\hat{e}$.

$$y_i = \beta_0 + \beta_1 x_{i1} + \dots + \beta_n x_{in} + \epsilon_i \tag{6.1}$$

$$\hat{y}_i = \hat{\beta}_0 + \hat{\beta}_1 x_{i1} + \dots + \hat{\beta}_n x_{in} + \hat{e}_i \tag{6.2}$$

In 6.1 and 6.2, x represents the value(s) of our predictor variable (the x-axis in Fig. 6.1). The slope ($\hat{\beta}$) of our line tells us how much $\hat{y}$ changes as we change one unit of x. In other words, $\hat{\beta}$ is the effect size of x—our only predictor (explanatory variable) here.[4] Crucially, when we run a model, our objective is to estimate both β_0 and β, that is, we want to fit a line to our data. In the sections that follow, we will explore multiple examples of linear models. We will see how to run them using R, how to interpret their results, and how to use them to predict patterns in our data. We will also discuss how you can present and report the results of a linear model.

6.2 Examples and Interpretation

Thus far, we have discussed data wrangling (dataPrepLinearModels.R) and data visualization (plotsLinearModels.R)—the focus of Part II. These are often the first two steps involved in data analysis. Let's now turn to the third step, the actual statistical analysis.

Our main research question here is simple: does feedback type affect participants' scores in our data, longFeedback? We can also ask whether other variables in the data affect scores. Are scores constant across tasks? Does it matter if participants spend more or fewer hours per week studying English? How about an effect of native language?

6.2.1 Does Hours *Affect Scores?*

The first thing we should do is plot the data to actually see what's going on. To answer our question (*does* Hours *affect scores?*), we want to have Hours on our x-axis and Score on our y-axis. Because both variables are continuous, a scatter plot is appropriate here. Our figure will therefore resemble Figs. 3.3 and 6.1—except for its aesthetics, since our data points are larger and have some transparency now, given that we have too many points in the figure.

Fig. 6.2 shows a scatter plot and a trend line. The gray dotted and dashed lines are there for illustrative purposes, to help with our discussion later—a typical scatter plot should only display a trend line in it. If you had to draw a line through the data such that it represented the trend observed, the solid line in the figure is the line you'd draw, that is, the best fit. We can see here that the line is inclined—it has a *positive slope* (much like the line in Fig. 6.1). It seems that the more a participant studies every week, the better his/her scores will be *on average*. You may remember that we have already created such a figure back in chapter 3 (Fig. 3.3). This time, however, we have no facets for native language because we want our figure to be aligned with our analysis that follows: we are examining only one predictor here, namely, Hours—nothing else.

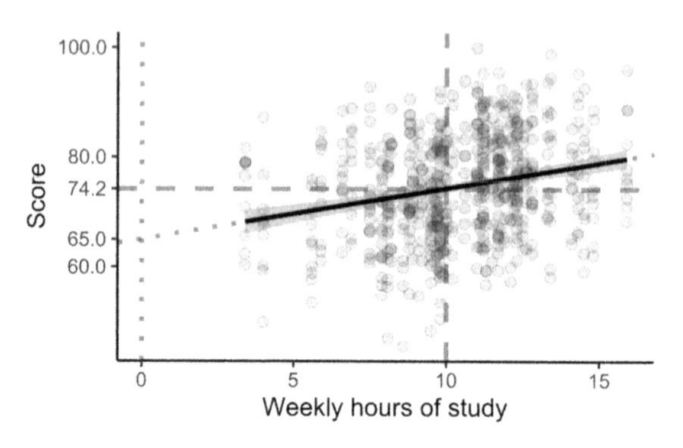

FIGURE 6.2 Participants' Scores by Weekly Hours of Study

We are now ready to run our model, since we already have a sense of what we see in the data: we want to check whether the effect of Hours on Score (the slope of our trend line) is statistically real. To run a linear model in R we use the function lm().

To answer our question earlier, we run lm(Score ~ Hours, data = longFeedback)[5]—note that our model mirrors Fig. 6.2: in both cases, we want to predict scores based on the number of hours of study (see §2.6.2.1). In code block 25, which should go at the top of our linearModels.R script, line 10 runs our model and assigns it to a variable, fit_lm1. This will be our first model fit to the data: we are modeling each participant's score as a function of Hours in our hypothetical study. Line 11 displays the results using the display() function from the arm package (Gelman and Su 2018). Line 12 is the most common way to print the output of our model—summary(). Finally, because we don't get confidence intervals by default in our output, line 13 prints them for us.

Let's inspect the most important information in the main output of our model, pasted in code block 25 as a comment (lines 15–25). The first column in our output, Estimate, contains the *coefficients* in our model. These are our effect sizes ($\hat{\beta}_0$ and $\hat{\beta}$). Our intercept, $\hat{\beta}_0 = 65.09$, represents the predicted score of a participant when he/she studies zero hours per week, that is, Hours = 0. This should make sense if you go back to 6.1 or 6.2: $\hat{\beta}_0 + \hat{\beta} \times 0 = \hat{\beta}_0$.

The 95% confidence intervals of the estimates are listed in lines 27–30 in code block 25 and could also be reported.

Line 19 tells us the effect of Hours: $\hat{\beta} = 0.92$. This means that for every additional hour of study (per week), a participant's score is predicted to increase 0.92 points. In other words, if a student studies 0 hours per week, his/her predicted score is approximately 65 ($\hat{\beta}_0$). If that same student studied 10 hours per week, then his/her predicted score would be 9.2 points higher:

$\hat{\beta}_0 + \hat{\beta}x \rightarrow 65 + 10 \cdot 0.92 = 74.2$. You can see this in Fig. 6.2: it's where the dashed lines cross.

Chances are you will only look at two columns in the output of a model when you use the summary() function: the estimate column, which we have just discussed, and the last column, where you can find the *p*-values for the estimates. In R, *p*-values are given using scientific notation. For example, our intercept has a *p*-value of 2e-16, that is, $2 \cdot 10^{-16}$ (2 preceded by sixteen zeros). This is clearly below 0.05, our alpha value. Indeed, we can see that both the intercept and the effect of Hours are significant. Let's understand what that means first and then examine the other columns in our output in lines 15–25.

What does it mean for an intercept to be significant? What's the null hypothesis here? The null hypothesis is that the intercept is *zero*—H_0: $\beta_0 = 0$. Remember: H_0 is always based on the assumption that an estimate is zero, whether it's the intercept or the estimate of a predictor variable. In other words, it assumes that the mean score for learners who study zero hours per week is, well, *zero*.

R code

```
1  # Remember to add this code block to linearModels.R
2  
3  source("dataPrepLinearModels.R")
4  # install.packages("arm")          # If you haven't installed this package yet
5  library(arm)                       # To generate a cleaner output: display()
6  
7  head(longFeedback)
8  
9  # Simple linear regression: continuous predictor:
10 fit_lm1 = lm(Score ~ Hours, data = longFeedback):
11 display(fit_lm1)
12 summary(fit_lm1)
13 confint(fit_lm1)
14 
15 # Output using summary(fit_lm1)
16 # Coefficients:
17 #                Estimate Std. Error t value Pr(>|t|)
18 #    (Intercept)  65.0861     1.5876  40.997  < 2e-16 ***
19 #    Hours         0.9227     0.1493   6.181 1.18e-09 ***
20 #    ---
21 #    Signif. codes:  0 '***' 0.001 '**' 0.01 '*' 0.05 '.' 0.1 ' ' 1
22 #
23 # Residual standard error: 9.296 on 598 degrees of freedom
24 # Multiple R-squared:  0.06005,        Adjusted R-squared:  0.05848
25 # F-statistic:  38.2 on 1 and 598 DF,  p-value: 1.18e-09
26 
27 # Confidence intervals using confint(fit_lm1):
28 #                  2.5 %    97.5 %
29 # (Intercept) 61.9681988 68.204066
30 # Hours        0.6295188  1.215872
```

CODE BLOCK 25 Simple Linear Regression and Output with Estimates: Score ~ Hours

But the estimate for the intercept is $\hat{\beta}_0 = 65$—clearly *not* zero. Thus, it shouldn't be that surprising that the intercept is significant. Think about it: how likely is it that the participants' scores should be zero if they studied zero hours per week? Not very likely. After all, even the worst scores would probably not be zero. We therefore have to reject H_0 here. To refresh our memories: the *p*-value here (< 0.001) represents the probability of finding data at least as extreme as the data we have assuming that the null hypothesis is true, so it's the probability of finding $\hat{\beta}_0 = 65$ when we assume it's actually zero ($\beta_0 = 0$). As we can see, the probability is exceptionally low—practically zero.

The *p*-value for our predictor, Hours, is also significant. The null hypothesis is again that $\beta = 0$, which in practical terms means that we assume the trend line in Fig. 6.2 is flat. The fact that $\hat{\beta}$ is positive and significant here tells us that the slope of our line is above zero, that is, not flat. Therefore, the number of hours a learner studies per week statistically affects his/her scores, and we again have to reject H_0. How much does Hours affect Score? 0.92 points for every weekly hour of study, our $\hat{\beta}$. These are all the estimates in our model, of course, so now let's take a look at the other columns in our output.

It turns out that all four columns in our model's output are connected. For example, if you divide the estimate by the standard error, the result will be the *t*-value[6] for the estimate. So our third column is simply our first column divided by our second column. And $|t|$-values above 1.96 will be significant (assuming $\alpha = 0.05$). For Hours, for example, our *t*-value is 6.2. Therefore, we already know that this predictor has a significant effect even without looking at the *p*-value column. Furthermore, because we know that the result here is statistically significant, we also know that the confidence interval for our predictor doesn't include zero, by definition. This is the reason that if you run display (fit_lm1), line 11 (from the arm package), the output only shows you estimates and standard errors: that's all you really need. But bear in mind that even though you only need these two columns, the vast majority of people in our field will still want to see *p*-values—and most journals will likely require them.

If you remember our brief review of standard errors back in §1.3.4, you may remember that we can manually calculate any 95% confidence interval ourselves by using the estimate and its standard error, so technically we don't even need the confidence intervals in our output. Take the estimated effect of Hours, $\hat{\beta} = 0.92$, and its standard error, 0.15. To manually calculate the lower and upper bounds of the 95% confidence interval, we subtract and add from the estimate its standard error times 1.96: 95% CI $= 0.92 \pm (1.96 \cdot 0.15)$. More generally, then, 95% CI $= \hat{\beta} \pm (1.96 \cdot SE)$. Alternatively, we can (and should) use the confint() function in R to have confidence intervals calculated for us—note that these will *not* be identical to the intervals manually calculated, as confint() uses a profile method to calculate intervals, which requires more computation.[7] The take-home message is this: R will tell us everything we need if we run summary() and confint() on a model fit. But you should know that most of the information here can be derived from *two numbers*, namely, the estimate and its standard error (for the intercept and any other predictor you may have in the model).

Finally, line 24 lists our R^2, which here is approximately 0.06.[8] This basically means that hours of study (as a predictor) explains about 6% of the variation that we observe in the scores in our data—if you think that this is a very low R^2, I agree. The last line in our output, line 25, simply tells us that the linear regression as a whole is significant—this is like running an ANOVA and finding a significant *F*-statistic, in case you were wondering.

Let's go back once more to Fig. 6.2 to wrap up our discussion on the model. The intercept of our model, $\hat{\beta}_0 \approx 65$ is the predicted score when Hours = 0. In our figure, the vertical dotted line represents Hours = 0 on the x-axis. If you look at the figure, the intercept is the point where the two dotted lines meet, which is indeed when $\hat{y} \approx 65$. The effect of Hours, in turn, is how much Score changes for each unit of Hours. Say we compare again 0 and 10 hours per week. Given the effect of Hours, $\hat{\beta} = 0.92$, we expect an increase of 9.2 points, so if you draw a vertical line at $x = 10$ in the figure, you should hit the trend line at $y = 65 + 9.2 = 74.2$—and that's where the dashed lines meet in Fig. 6.2.

REPORTING RESULTS

A linear model shows that weekly hours of study has a significant effect on learners' scores ($\hat{\beta} = 0.92, 95\%\,\text{CI} = [0.63, 1.22], p < 0.001; R^2 = 0.06$). These results indicate that one additional weekly hour of study had an average positive impact of 0.92 point on learners' scores.

6.2.2 *Does* Feedback *Affect Scores?*

Let's now turn to our main question. If we want to examine whether feedback affects scores, we should first create a figure where we plot both variables. We will then run another simple linear model that predicts a participant's score as a function of feedback type. We can represent that relationship as Score ~ Feedback.

One way to phrase our question here would be: "Is there a difference in score means between the group that received recast and the group that received explicit feedback?" A better question, however, would be "What is the effect of feedback on participants' scores?" While the first question can be answered

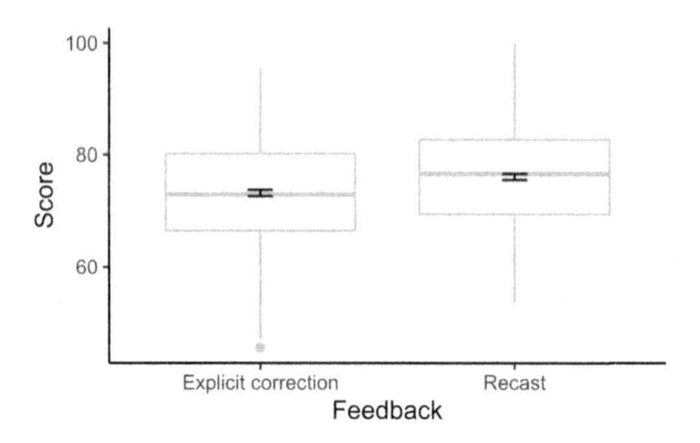

FIGURE 6.3 Participants' Scores Across Both Feedback Groups

with a simple "yes" if we find significant results, the second question focuses on the size of the effect—a much more relevant aspect of our hypothetical study here. Think about it this way: something can be statistically significant but practically meaningless. Our focus in linear models should be first on the effect, not on whether there's a difference. That's what we did earlier for our continuous predictor (Hours), and that's exactly how we will approach our next model—only now with a categorical predictor, Feedback.

Fig. 6.3 shows how scores differ for both feedback groups. We can see that the mean score for the recast group is slightly higher than the mean score for the explicit correction group. If we were to draw a line between the two means (inside the error bars), the line would be positively inclined (from left to right), that is, we would have a *positive slope* between the two levels of Feedback. In other words, even though here we have a categorical variable on the x-axis (cf. Hours in Fig. 6.2), we can still think of our model the same way we did for the discussion in §6.2.1.

To answer our main question earlier (*does* Feedback *affect scores?*) we can run lm(Score ~ Feedback, data = longFeedback)—here again our model mirrors Fig. 6.3: in both cases we want to predict scores based on the two feedback groups in question (see §2.6.2.1). In code block 26, line 2 runs our model and assigns it to a variable, fit_lm2. This will be our second model fit to the data: we are modeling each participant's score as a function of which feedback his/her group received in our hypothetical study. Lines 3–5 should be familiar from code block 25—this time, we will use display()[9] to generate our output, so you can decide which one you prefer (summary() or display()).

Let's inspect the most important information in the output of our model, pasted in code block 26 as a comment (lines 7–14). Before, when we ran fit_lm1, the intercept meant "the predicted score when Hours = 0". What does it mean now? Essentially, the same thing: the predicted score when

TABLE 6.1 Example of Contrast Coding

Feedback		*Recast*
Explicit correction		0
Explicit correction		0
Recast	→	1
Recast		1
Explicit correction		0
Recast		1
...		...

Feedback = 0. But what's zero here? R will automatically order the levels of Feedback alphabetically: Explicit correction will be our reference level, and Recast will be coded as either 0 or 1 accordingly. This type of **contrast coding** is known as *treatment* coding, or *dummy* coding—see Table 6.1 and Table E.1 in Appendix E for an example of dummy coding for a factor with three levels. Therefore, our intercept here represents the group of participants who received explicit correction as feedback. You can see that because here Feedback only has two levels, and Recast is listed in line 11.[10]

We can mathematically represent our linear model as $\hat{y} = \hat{\beta}_0 + \hat{\beta}_{\text{recast}} \cdot x$, where $x = 0$ (explicit correction) or $x = 1$ (recast). Note that because we only have two levels in Feedback, we only need one $\hat{\beta}$, since we can use 0 or 1—that is, only one new column in Table 6.1 is sufficient to capture two levels. If Feedback had three levels, we would have *two* $\hat{\beta}$s $(n - 1)$—that is, two new columns would be needed in Table 6.1.

Imagine we had three types of feedback in the data, namely, Explicit correction, Recast, and Peer (for student peer feedback). Our model would be represented as follows: $\hat{y} = \hat{\beta}_0 + \hat{\beta}_{\text{recast}} \cdot x + \hat{\beta}_{\text{peer}} \cdot x$—notice that we only have two $\hat{\beta}$s. For a participant in the Peer group, we'd have $\hat{y} = \hat{\beta}_0 + \cancel{\hat{\beta}_{\text{recast}} \cdot 0} + \hat{\beta}_{\text{peer}} \cdot 1$. For a participant in the Recast group, we'd have $\hat{y} = \hat{\beta}_0 + \hat{\beta}_{\text{recast}} \cdot 1 + \cancel{\hat{\beta}_{\text{peer}} \cdot 0}$. And for a participant in the Explicit correction group, we'd have $\hat{y} = \hat{\beta}_0 + \cancel{\hat{\beta}_{\text{recast}} \cdot 0} + \cancel{\hat{\beta}_{\text{peer}} \cdot 0}$. In summary, a categorical predictor will be transformed into 0s and 1s in our models (via **contrast**), and our intercept will represent one of the levels (by default, the first level in alphabetical order).

It should now be clear that the intercept here tells us the mean score for the explicit correction group ($\bar{x} = 73.2$), and line 11 tells us that the recast group had an average score of 2.9 points *higher* than that, given the positive $\hat{\beta}$. In other words, the mean score for explicit correction is 73.2, and the mean score for recast is 76.1—look back at Fig. 6.3 and you will be able to connect the statistical results with the trend shown in the plot.

How can we tell whether our intercept is significant if there are no p-values in our output shown in code block 26? Remember: all we need are estimates and their standard errors, so the minimalism of the output here shouldn't be a problem. If we divide $\hat{\beta}_0$ by its standard error, we will clearly get a number that is higher than 1.96. Therefore, given such an extreme t-value ($73.17 \div 0.55 = 133.04$), we will have a highly significant p-value. The null hypothesis is that the intercept is *zero* ($H_0 : \beta_0 = 0$)—this is always the null hypothesis. In other words, it assumes that the mean score for the explicit instruction group is zero. We have to reject H_0 here because our estimate is statistically not zero ($\hat{\beta} = 73.2$).

The same can be said about the effect of Feedback: $2.89 \div 0.77 = 3.75$. We therefore conclude that the effect of feedback is also significant. This is like saying that the difference between the two groups (= 2.9) is statistically significant. In summary: because our standard errors here are less than 1, and because our estimates are both greater than 2, both $|t|$-values will necessarily be greater than 1.96. Naturally, you can always run summary() on your model fit if you want to have p-values explicitly calculated for you.

Our R^2 for this model is 0.02—so a model with Feedback explains less variation in scores relative to a model with Hours as its predictor. Needless to say, 0.02 is a low R^2, but the question you should ask yourself is how much you

R code

```
1  # Simple linear regression: categorical predictor
2  fit_lm2 = lm(Score ~ Feedback, data = longFeedback)
3  display(fit_lm2)
4  summary(fit_lm2)
5  confint(fit_lm2)
6
7  # Output using display(fit_lm2):
8  # lm(formula = Score ~ Feedback, data = longFeedback)
9  #                coef.est  coef.se
10 # (Intercept)     73.17     0.55
11 # FeedbackRecast   2.89     0.77
12 # ---
13 # n = 600, k = 2
14 # residual sd = 9.48, R-Squared = 0.02
15
16 # Confidence intervals using confint(fit_lm2):
17 #                       2.5 %    97.5 %
18 #     (Intercept)    72.096209 74.245791
19 # FeedbackRecast      1.367016  4.406984
```

CODE BLOCK 26 Simple Linear Regression and Output with Estimates: Score ~ Feedback

care about this particular metric in your research. If your main objective is to show that feedback *has an effect*, then a low R^2 is not your top priority. If, on the other hand, you want to argue that feedback is a powerful predictor to explain learners' scores, then you shouldn't be too excited with such a low R^2. Ultimately, a simple model with a single predictor is unlikely to explain much of the data when it comes to human behavior. There are so many factors at play when we deal with language learning that a realistic model will necessarily have to be more complex than the model we are examining here.

As you can see, the effect size here tells us the difference it makes to go from explicit correction to recast. That the effect size is small is not inconsistent with the literature: Russell and Spada (2006, p. 153), for example, conclude that "[w]hile the results from this meta-analysis indicate that CF [corrective feedback] is useful overall, it was not possible to determine whether particular CF variables make a difference".

Finally, bear in mind that both models we have run thus far have essentially the same underlying structure: $\hat{y}_i = \hat{\beta}_0 + \hat{\beta}x_i + \hat{e}_i$, as per 6.1 or 6.2. Next, we will run a model with two predictors, that is, two $\hat{\beta}$s.

> **REPORTING RESULTS**
>
> A linear model shows that feedback has a significant effect on learners' scores ($\hat{\beta} = 2.9, 95\%\,\text{CI} = [1.4, 4.4], p < 0.001; R^2 = 0.02$).[11] Learners who received recast as feedback had an average score 2.9 points higher than those who received explicit correction.

6.2.3 *Do* Feedback *and* Hours *Affect Scores?*

Most of the time you will not run simple linear models. Realistic models have *multiple* predictors because we typically assume that several factors are driving a particular effect—especially when it comes to learning languages. Our question here can therefore be phrased as follows: "once we take into account the effect of Hours, what's the effect of Feedback (and vice versa)?" It could be the case that once we include Hours in our model, the effect of Feedback goes away. Therefore, the best we can do is have a model that includes *both* variables in it, that is, a multiple linear regression.

Fig. 6.4 plots our three variables: our response variable on the y-axis (Score), Hours on the x-axis, and Feedback using facets.[12] Our model now is specified as `lm(Score ~ Feedback + Hours, data = longFeedback)`. Let's examine the model output in code block 27 using `summary()`.

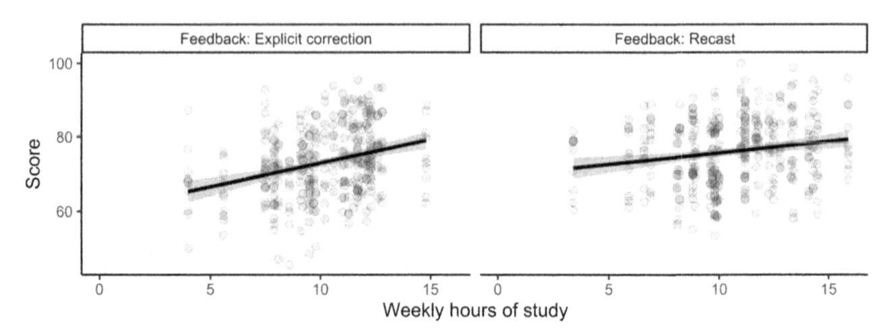

FIGURE 6.4 Participants' Scores by Weekly Hours of Study and by Feedback Type

Our intercept ($\hat{\beta}_0$) is 64. What does that mean now that we have two predictors? Well, it means exactly the same thing as before: $\hat{\beta}_0$ here represents the predicted score when our other terms ($\hat{\beta}$) are zero: Feedback = Explicit correction and Hours = 0. This should make sense—we're merely combining the interpretation of the intercepts of the two models discussed earlier.

The estimate for Feedback = Recast, presented in line 11, represents the change in score when we go from the explicit correction group to the recast group. Likewise, the estimate for Hours represents the change in score for every hour a participant studies every week. All estimates are statistically significant ($p < 0.001$), which should not be too surprising given what we have discussed so far.

We can represent our model mathematically as $\hat{y}_i = \hat{\beta}_0 + \hat{\beta}_1 x_{i1} + \hat{\beta}_2 x_{i2} + \hat{e}_i$. Here, we have one intercept ($\hat{\beta}_0$) and two predictors ($\hat{\beta}_1$ and $\hat{\beta}_2$). Recall that $\hat{y}_i$ represents the predicted response (i.e., the score) of the ith participant. Let's say we have a participant who is in the recast group and who studies English for 7 hours every week. To estimate the score of said participant, we can replace the variables in our model as follows: $\hat{y}_i = 64.2 + 2.55 \cdot 1 + 0.88 \cdot 7 = 72.91$. Notice that our β_1 here represents Feedback, and it's set to 1 if Feedback = Recast and to 0 if Feedback = Explicit correction. Hopefully now it's clear why the intercept is our predicted response *when all other variables are set to zero*: if we choose a hypothetical participant who studies zero hours per week and who is in the explicit correction group, his/her predicted score will be: $\hat{y}_i = 64.2 + 2.55 \times 0 + 0.88 \times 0 = 64.2$, and that is exactly $\hat{\beta}_0$.

REPORTING RESULTS

A linear model shows that both feedback (recast) and weekly hours of study have a significant positive effect on learners' scores. Learners in the recast group had higher mean scores than learners in the explicit correction group ($\hat{\beta} = 2.55, 95\%$ CI = [1.06, 4.03], $p < 0.001$). In addition, learners with a higher number of weekly study hours also had higher mean scores ($\hat{\beta} = 0.88, 95\%$ CI = [0.59, 1.18], $p < 0.0001$)

Note that our (adjusted) R^2 is now above 0.07.[13] This is still a low number: by using Feedback and Hours we can only explain a little over 7% of the variation in scores in the data.

Here's a question: looking at the estimates in code block 27, can we say that feedback is more important than hours of study, given that their coefficients are $\hat{\beta} = 2.55$ and $\hat{\beta} = 0.88$, respectively? The answer is *no*, we can't say that. We have to resist the temptation to directly compare these two numbers: after all, our variables are based on completely different scales. While Feedback can be either 0 or 1, Hours can go from 0 to 10 (and higher). Therefore, these two estimates are *not* comparable right now. We will discuss how to make them comparable later in this chapter (§6.3.2). For now, we can't say which one has a stronger effect on scores.[14]

R code
```
# Multiple linear regression: categorical and continuous predictor
fit_lm3 = lm(Score ~ Feedback + Hours, data = longFeedback)
display(fit_lm3)
summary(fit_lm3)
confint(fit_lm3)

# Output excerpt using summary(fit_lm3)
# Coefficients:
#                   Estimate Std. Error t value Pr(>|t|)
#   (Intercept)      64.2067     1.5955  40.243  < 2e-16 ***
#   FeedbackRecast    2.5449     0.7547   3.372 0.000794 ***
#   Hours             0.8846     0.1484   5.960 4.32e-09 ***
#   ---
#   Signif. codes:  0 '***' 0.001 '**' 0.01 '*' 0.05 '.' 0.1 ' ' 1
#
# Residual standard error: 9.217 on 597 degrees of freedom
# Multiple R-squared:  0.07762,        Adjusted R-squared:  0.07453
# F-statistic: 25.12 on 2 and 597 DF,  p-value: 3.356e-11

# Confidence intervals using confint(fit_lm2):
#                       2.5 %    97.5 %
# (Intercept)     61.0733247 67.340123
# FeedbackRecast   1.0627199  4.027164
# Hours            0.5931205  1.176145
```

CODE BLOCK 27 Multiple Linear Regression: Score ~ Feedback + Hours

If you look back at Fig. 6.4, you will notice that the two trend lines in the figure seem to be slightly different. More specifically, the line for the explicit correction group is more inclined than the line for the recast group. If that's true, what does it mean? It means that the effect of Hours is *stronger* for the participants in the explicit correction group. If statistically real, this effect would tell us that the two variables in question *interact*. In other words, the effect of Hours depends on whether a learner is in the explicit correction group or in the recast group. This is what we will examine next.

6.2.4 *Do* Feedback *and* Hours *Interact?*

Fig. 6.5 is essentially the same as Fig. 6.4, but here we use a single facet and use linetype to differentiate the two types of Feedback. The disadvantage is that we can no longer differentiate the data points from the two feedback groups, as now they're all in a single facet.[15] But the advantage of Fig. 6.5 is that we can more easily see that the two trend lines are *not* parallel. When lines representing a categorical variable along a continuous variable are not parallel in a scatter plot, chances are there's an interaction between the variable on the x-axis (Hours) and said categorical variable (Feedback). Here, we can see that the slope of the solid line is more inclined than the slope of the dashed line, suggesting that the effect of Hours on Score is *stronger* for the explicit correction group. We will verify whether that trend is statistically real next by running a linear model with an interaction between the two predictors in question.

To run a model with an interaction between two variables, we use an asterisk "*". Thus, our model here is defined as Score ~ Feedback * Hours—line 3 in code block 28. When we run such a model, we are actually checking for an effect of Feedback, an effect of Hours, and an effect of both variables together. Indeed, we could define the model as Score ~ Feedback + Hours +

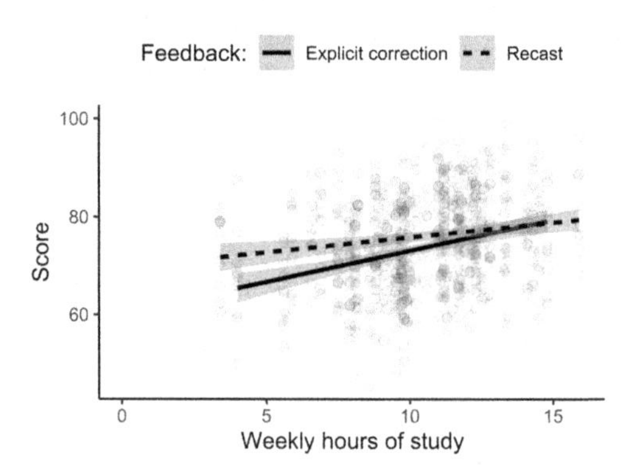

FIGURE 6.5 A Figure Showing How Feedback and Hours may Interact

Feedback:Hours. The colon here represents an interaction, while the asterisk represents "main effects *plus* an interaction".

Mathematically, we can define our model as $\hat{y}_i = \hat{\beta}_0 + \hat{\beta}_1 x_{i1} + \hat{\beta}_2 x_{i2} + \hat{\beta}_3 x_{i1} x_{i2} + \hat{e}_i$. Note that x_1 represents the value of Feedback for participant i, and x_2 represents the value of Hours for participant i. Thus, we have both main effects and an interaction. As usual, $\hat{\beta}_0$ represents our estimated intercept (the predicted score of a participant when Feedback = 0 (= Explicit correction), and when Hours = 0. The estimate for $\hat{\beta}_0$ can be found in code block 28. Let's now interpret the results of our model, which will tell us what our estimates ($\hat{\beta}$) are.

As we go over the effects of our variables in fit_lm4, you will notice that interpreting a model with interactions is slightly different than interpreting a model without interactions. First things first: all three estimates and our intercept are significant. This also means that the interaction that we suspected existed is statistically real. As usual, we can see all our estimates in the Estimate column in the output shown in code block 28 (lines 8–20). It's true that we could simply say that our model shows a significant effect of Feedback, Hours, as well as a significant interaction of both variables. That is what most people will say about results such as these. But saying that doesn't mean that we actually understand what the results mean—especially when it comes to the interaction term. So let's dig a little deeper here, since this is the most complex model we have run thus far.

INTERCEPT. The intercept represents the predicted score when Feedback = 0 (explicit correction) and Hours = 0. A participant who has zero hours of study per week and is in the explicit correction group is therefore predicted to have an average score of 60 points (our $\hat{\beta}_0$). This participant, of course, doesn't exist in our data, but it's still a prediction of our model. Going back to the mathematical structure of our model, we can easily replace the numbers here (ignoring our error term, $\hat{e}$): $\hat{y}_i = \hat{\beta}_0 + \hat{\beta}_1 x_{i1} + \hat{\beta}_2 x_{i2} + \hat{\beta}_3 x_{i1} x_{i2} \rightarrow 60.28 + \cancel{9.28 \cdot 0} + \cancel{1.27 \cdot 0} + \cancel{(-0.65 \cdot 0 \cdot 0)} = 60.28$.

FEEDBACK. The result in line 12, FeedbackRecast, tells us the effect of being in the recast group (this should be familiar at this point given our discussion about fit_lm2). The important thing to remember here is that for us to interpret *one* estimate, we need to keep the other estimates constant. Here's what that means: a participant in the recast group has a predicted score of 9.3 points higher than a participant in the explicit correction group *assuming that both participants have zero hours of weekly study*—it will be clear why that's the case once we examine our interaction, but if you look back at Fig. 6.5, the crossing of our lines should give you a hint.

We can again replace the coefficients in our model with our actual estimates for this example. A participant in the recast group will have the following predicted score: $\hat{y}_i = \hat{\beta}_0 + \hat{\beta}_1 x_{i1} + \hat{\beta}_2 x_{i2} + \hat{\beta}_3 x_{i1} x_{i2} \rightarrow 60.28 +$

R code

```
# Interaction linear regression: categorical and continuous predictor
# Note the * in the model specification below:
fit_lm4 = lm(Score ~ Feedback * Hours, data = longFeedback)
display(fit_lm4)
summary(fit_lm4)
round(confint(fit_lm4), digits = 2) # if you wish to round CIs (see lines 22-27)

# Output excerpt using summary(fit_lm4)
# Coefficients:
#                        Estimate Std. Error t value Pr(>|t|)
#   (Intercept)           60.2775     2.4074  25.039  < 2e-16 ***
#   FeedbackRecast         9.2825     3.1888   2.911  0.00374 **
#   Hours                  1.2724     0.2317   5.491 5.93e-08 ***
#   FeedbackRecast:Hours  -0.6547     0.3011  -2.174  0.03008 *
#   ---
#   Signif. codes:  0 '***' 0.001 '**' 0.01 '*' 0.05 '.' 0.1 ' ' 1
#
# Residual standard error: 9.188 on 596 degrees of freedom
# Multiple R-squared:  0.08488,      Adjusted R-squared:  0.08027
# F-statistic: 18.43 on 3 and 596 DF,  p-value: 1.918e-11
#
# Confidence intervals using confint(fit_lm2):
#                       2.5 % 97.5 %
# (Intercept)           55.55  65.01
# FeedbackRecast         3.02  15.55
# Hours                  0.82   1.73
# FeedbackRecast:Hours -1.25  -0.06
```

CODE BLOCK 28 Modeling an Interaction: Score ~ Feedback * Hours

$9.28 \cdot 1 + \xcancel{1.27 \cdot 0} + \xcancel{(-0.65 \cdot 1 \cdot 0)} \rightarrow 60.28 + 9.28 = 69.56$. And a participant in the explicit correction group will have the following predicted score: $\hat{y}_i = \hat{\beta}_0 + \hat{\beta}_1 x_{i1} + \hat{\beta}_2 x_{i2} + \hat{\beta}_3 x_{i1} x_{i2} \rightarrow 60.28 + \xcancel{9.28 \cdot 0} + \xcancel{1.27 \cdot 0} + \xcancel{(-0.65 \cdot 1 \cdot 0)} = 60.28$—yes, this is exactly what we calculated for our intercept earlier. Note that both examples assume Hours $= 0$ and that the difference in predicted score for these two participants is *exactly* the effect size of Feedback: $\hat{\beta} = 9.28$.

HOURS. The effect of hours ($\hat{\beta} = 1.27$) tells us that for every extra weekly hour of study, a participant's score is predicted to increase by 1.27 points *assuming that the participant is in the explicit correction group*. You can replace the coefficients with the estimates as we did for Feedback earlier.

FEEDBACKRECAST:HOURS. The last estimate we have in code block 28 represents our interaction term. The null hypothesis here is that there is no interaction between the two variables; that is, the lines in Fig 6.5 are parallel. The estimate here is negative ($\hat{\beta} = -0.65$), which means that the effect of Hours weakens once we go from the explicit correction group to the recast group. Does this make sense? Look back at Fig. 6.5 and you will see that the line representing the recast group is *less* inclined than the line representing the explicit correction group. I mentioned earlier that this difference in the figure suggested that the effect of hours was weaker for the recast group—and that's why we now have a negative estimate for the interaction in the model. Let's see that in action by considering two examples: participant A will be in the recast group and reports studying 5 hours per week. Participant B is in the explicit correction group and also reports studying 5 hours per week. The predicted score for both participants is calculated in 6.3.

$$
\begin{aligned}
\text{Participant A}: \quad & \hat{y}_A = \hat{\beta}_0 + \hat{\beta}_1 x_{1A} + \hat{\beta}_2 x_{2A} + \hat{\beta}_3 x_{1A} x_{2A} \\
& \hat{y}_A = 60.28 + 9.28 \cdot 1 + 1.27 \cdot 5 + (-0.65 \cdot 1 \cdot 5) \\
& \hat{y}_A = 72.65 \\
\text{Participant B}: \quad & \hat{y}_B = \hat{\beta}_0 + \hat{\beta}_1 x_{1B} + \hat{\beta}_2 x_{2B} + \hat{\beta}_3 x_{1B} x_{2B} \\
& \hat{y}_B = 60.28 + \xcancel{9.28 \cdot 0} + 1.27 \cdot 5 + \xcancel{(-0.65 \cdot 0 \cdot 5)} \\
& \hat{y}_B = 66.64
\end{aligned}
\tag{6.3}
$$

In our example here, participant A (who's in the recast group) is predicted to have a higher score than participant B (who's in the explicit correction group). Do you think that will always be the case? Look back at the lines in Fig. 6.5. The two trend lines actually cross when the number of weekly hours is approximately 15. As a result, given enough weekly hours, the predicted score of participant B would actually be better than that of participant A. If you calculate the predicted scores again, this time assuming Hours = 20, you will see that $\hat{y}_A = 81.91$ and $\hat{y}_B = 85.73$. Being able to inspect Fig. 6.5 is certainly helpful when it comes to understanding interactions in statistical models.

You may be wondering whether there's an easy way to predict the scores of a participant using our model. After all, you probably don't want to manually add the numbers to our regression as we did earlier. We can easily do that with R by using the function `predict()`. For example, if you want to predict the scores of participants A and B from earlier, you could run `predict(fit_lm4, newdata = tibble(Feedback = c("Recast", "Explicit correction"), Hours = c(20, 20)))`. The argument `newdata` requires that variable names and values match exactly what we have in the data being modeled. The function `predict()` basically takes a model fit (`fit_lm4` here) and some new data and returns the predicted $\hat{y}$, that is, a score for each of the two participants we are examining here. The nice thing about the `predict()` function is that we can have it predict scores for any number of participants simultaneously. Try playing around with the function to see how helpful it can be—you could, for instance, create a figure showing predicted scores for participants who do not necessarily exist in your dataset.

In summary, the interaction here tells us that the effect of Hours is weakened if you are in the recast group. That explains why participant B can surpass

TABLE 6.2 Statistical Table Showing Model Estimates

	Estimate ($\hat{\beta}$)	*Std. Error*	t-*value*	*p*-value
(Intercept)	60.28	2.41	25.04	< 0.001
Feedback: Recast	9.28	3.19	2.91	< 0.01
Hours	1.27	0.23	5.49	< 0.001
Feedback (Recast):Hours	−0.65	0.30	−2.17	< 0.05

participant A if they both study 20 hours per week. By inspecting Fig. 6.5, it's much easier to see this interaction in action. A typical table with the model estimates is provided in Table 6.2—the table merely copies the output in code block 28.[16] You'll notice that we have no confidence intervals in the table. However, recall that we can calculate the 95% confidence intervals using the standard errors. Whether you need to explicitly report confidence intervals will depend in part on your readership (and on the author guidelines of the journal of your choice). Finally, notice that the table has no vertical lines dividing the columns and that numeric columns are right-aligned. As a rule of thumb, your tables should follow a similar pattern.

REPORTING RESULTS

A linear model shows a significant effect of Feedback ($\hat{\beta} = 9.28$, $p < 0.01$) and Hours ($\hat{\beta} = 1.27, p < 0.0001$) as well as a significant interaction between the two variables ($\hat{\beta} = -0.65, p < 0.05$). The negative estimate for the interaction tells us that the effect of Hours is weaker for participants in the recast group. As a result, our model predicts that although participants in the recast group have higher scores on average, they can be surpassed by participants in the explicit correction group given enough weekly hours of study, as shown in the trends in Fig. 6.5.

The earlier discussion is way more detailed than what you need in reality. We normally don't explain what an interaction means: instead, as mentioned earlier, we simply say that said interaction is significant. This is in part because we assume that the reader will know what that means—you should certainly consider whether that assumption is appropriate.

Thus far, we have run four models. A natural question to ask is which one we should report. Clearly, we don't need all four: after all, they all examine the same question (which variables affect Score). So let's see how we can compare all four models.

6.3 Beyond the Basics

6.3.1 Comparing Models and Plotting Estimates

In this section we will use a function to compare the fit of different models in R. Once we choose the model with the best fit, we will then create a plot for the estimates. There are many different ways we can compare how good a model is relative to other models (assuming the models are run on the same dataset, of course). A quick way to do that is to use the anova() function in R. Before we actually compare our models, can you guess which of the four models discussed earlier offers the best fit given our data? How would you know that?

First, let's focus on our most complex model, fit_lm4. In that model, shown in Table 6.2, we see that both Feedback and Hours have significant effects on Score (and so does the interaction between the two variables). Now that we know that, you can already conclude that the other fits are likely *less* optimal: after all, they leave out effects that we now know are statistically significant. For example, fit_lm1 only considers Feedback. But we know that the variable Hours also matters. As a result, relative to fit_lm4 it shouldn't be surprising that fit_lm1 is inferior. The most reasonable comparison here would therefore be fit_lm3 and fit_lm4.

Let's inspect code block 29, where we run the anova() function to compare fit_lm3 and fit_lm4—this will print an "Analysis of Variance Table" in your console. Recall that the only difference between these two models is the interaction in fit_lm4. Therefore, by comparing these two models we are asking "how much do we get by adding an interaction?" At the end of line 11, we see a *p*-value below 0.05 and an asterisk. That indicates that model 2 (fit_lm4) is statistically better than model 1 (fit_lm3). The RSS column shows us the residual sum of squares, which you may recall is essentially a

R code

```
1  # Compare models:
2  anova(fit_lm3, fit_lm4)
3
4  # Output:
5  # Analysis of Variance Table
6  #
7  # Model 1: Score ~ Feedback + Hours
8  # Model 2: Score ~ Feedback * Hours
9  #   Res.Df   RSS Df Sum of Sq      F  Pr(>F)
10 # 1    597 50712
11 # 2    596 50313  1     399.1 4.7276 0.03008 *
12 #   ---
13 #   Signif. codes:  0 '***' 0.001 '**' 0.01 '*' 0.05 '.' 0.1 ' ' 1
```

CODE BLOCK 29 Comparing Models Using anova()

number telling us how much our model deviates from the actual data being modeled. Thus, the lower the RSS, the better. Model 2 has a lower RSS (50313 *vs.* 50712), which means it has a better fit relative to model 1. In sum, we can say that there is a significant effect of the interaction between Feedback and Hours ($F(1, 596) = 4.73$, $p \approx 0.03$).[17]

The earlier comparison shows that fit_lm4 offers the best fit of the data given the models we have discussed so far. We have already seen how to report the results from the model in text and in Table 6.2. Now let's see how to present model estimates using a plot. In our plot, we will have error bars that represent estimates and their respective 95% confidence intervals. The result is shown in Fig. 6.6—you can see how the figure is created in code blocks 30 and 31.

The *y*-axis in Fig. 6.6 lists the predictors in our model (fit_lm4)—following the same order that we have in Table 6.2. In the figure, we can easily see the estimates (they are in the center of the error bars) and the confidence intervals (the error bars themselves). The figure also prints the actual estimates under each error bar, but that is likely not necessary, given the *x*-axis. Being able to see confidence intervals in a figure can be more intuitive, since we can actually see how wide each interval is. Recall that 95% confidence intervals that cross (or include) zero mean that the estimate in question is not significant ($p > 0.05$). Here, none of the error bars cross zero on the *x*-axis—although the interaction comes close, so it's not surprising that its *p*-value in our model is closer to 0.05 in Table 6.2. We could remove the intercept from our figure to reduce the range on the *x*-axis and better visualize the predictors of interest—it's not uncommon for researchers to remove intercepts from statistical tables, especially if their meaning is not practically relevant.

Code blocks 30 and 31 show how Fig. 6.6 is created.[18] First, we have to prepare the data (code block 30) by gathering estimates and their confidence intervals. Only then can we actually create the figure (code block 31). All the lines of code in both code blocks show you how to create the figure *manually*, just in case you were wondering how this could be done—spend some time inspecting the lines of code that create lm4_effects, as some of it will

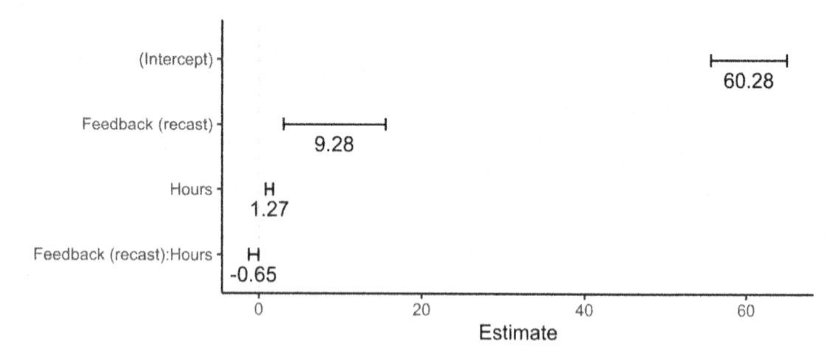

FIGURE 6.6 Plotting Estimates from a Linear Model

R code

```
# Remember to add this code block to plotsLinearModels.R
# For the code to work, you must have already run fit_lme4
source("dataPrepLinearModels.R")

# Plot estimates and confidence intervals for fit_lm4
# Prepare data:
lm4_effects = tibble(Predictor = c("(Intercept)",
                                    "Feedback (recast)",
                                    "Hours",
                                    "Feedback (recast):Hours"),
                     Estimate = NA,
                     l_CI = NA,
                     u_CI = NA)

# Add coefficients and confidence intervals (CIs):
lm4_effects = lm4_effects %>%
  mutate(Estimate = c(coef(fit_lm4)[[1]],  # Intercept
                      coef(fit_lm4)[[2]],  # Feedback
                      coef(fit_lm4)[[3]],  # Hours
                      coef(fit_lm4)[[4]]), # Interaction
         l_CI = c(confint(fit_lm4)[1:4]),  # lower CI
         u_CI = c(confint(fit_lm4)[5:8]))  # upper CI

# Visualize tibble:
lm4_effects
```

CODE BLOCK 30 Preparing the Data for Plotting Model Estimates and Confidence Intervals

be familiar. However, if you're in a hurry, you could simply install the sjPlot package, load it, and then run lines 22–23 in code block 31. Yes: with only two lines you produce a figure that takes around 30 lines of code to produce manually (!). Naturally, the easy way will not produce exactly the same figure, but it will be essentially what you see in Fig. 6.6.

We have seen earlier that we can present the results of a statistical model using a table or a figure. We could also have the results in the body of the text and omit any tables of figures, but that's probably the worst option. Tables and figures are much better at presenting results in an organized way. The question you should ask yourself now is which option best suits your taste, needs, and readership: a table such as Table 6.2 or a figure such as Fig. 6.6. The vast majority of papers show model estimates in tables, not figures. However, in many cases figures will certainly provide a more intuitive way to discuss the results of your model. Indeed, much like box plots are underused in data visualization in our field, the same can be said for figures to represent model estimates, unfortunately.

R code
```
# Plot estimates and confidence intervals for fit_lm4 (add to plotsLinearModels.R)
# Plot estimates and 95% confidence intervals:
ggplot(data = lm4_effects, aes(x = Predictor, y = Estimate)) +
  geom_errorbar(aes(ymin = l_CI, ymax = u_CI), width = 0.2) +
  coord_flip() +
  theme_classic() +
  scale_x_discrete(limits = c("Feedback (recast):Hours",
                              "Hours",
                              "Feedback (recast)",
                              "(Intercept)")) +
  geom_text(aes(label = round(Estimate, digits = 2)),
            position = position_nudge(x = -0.3, y = 0)) +
  geom_hline(yintercept = 0, linetype = "dashed", alpha = 0.1) +
  labs(x = NULL)

# Save plot in figures folder:
# ggsave(file = "figures/model-estimates.jpg", width = 6, height = 2.5, dpi = 1000)

# Alternatively, install sjPlot package:
# library(sjPlot)
#
# plot_model(fit_lm4, show.intercept = TRUE) +
#   theme_classic()
```

CODE BLOCK 31 Plotting Model Estimates and Confidence Intervals

6.3.2 Scaling Variables

While discussing our model's estimates back in code block 27, I mentioned that we should avoid comparing different estimates in a single model. This is generally true for any model with multiple predictors: different predictors will often use different scales, so comparing estimates across predictors is not a reliable method to assess the relative impact of each predictor on the response variable.

Let's go back to our fit_lm4 model that contains Feedback, Hours, as well as the interaction between the two variables. Fig. 6.6 tells us that the effect size of Feedback is 9.28, while the effect size of Hours is 1.27. Intuitively, many people will jump to the conclusion that Feedback is more than seven times as strong or as important as Hours in predicting a learner's score. This is, however, not correct, as Feedback is a binary variable that takes only 0 or 1 as its values, while Hours is a continuous variable that can take any positive number as its value. In other words, the effect size we see here represents different things for these two variables: for Feedback, $\hat{\beta}$ represents the effect of going from 0 to 1—from one extreme to another extreme of the possible values Feedback can take. For Hours, on the other hand, the $\hat{\beta}$ simply represents the effect of *1* unit along the continuum of hours.

In this section, we will see how we can rescale all the variables in our model. By doing that, we will be able to directly compare the estimates we have been discussing. The disadvantage of scaling our variables is that our interpretation of the results becomes a little less intuitive to some people. But what does it mean to rescale a variable?

To rescale variable X, we can subtract from all of its values the mean of X ($\bar{x}$) and divide it by two standard deviations of X (s).[19] The result is that our new mean will be 0 (so the variable is centered), and each unit represents 2 standard deviations (so the variable is standardized). Indeed, for any rescaled variable, its new mean will be 0, and its standard deviation will be 0.5. Read the previous sentence very carefully again. For example, the mean of Hours is currently 10.3, and its standard deviation is 2.5. All the values that are currently below 10.3 will be negative once we rescale Hours—which is intuitive insofar as they are *below average*. Conversely, all the values that are currently above 10.3 will be positive once we rescale Hours, that is, they are *above* average.

You may be wondering how we can calculate the mean of Feedback if that's not even a numeric variable. As discussed in our previous models, a categorical variable such as Feedback will be processed as a numeric variable that has values 0 or 1: Explicit correction = 0 and Recast = 1. Therefore, we can easily calculate the mean and standard deviation of Feedback. The mean of a binary variable will always be 0.5, and its standard deviation will be 0.5 too.

Once we rescale both Feedback and Hours, both variables will have a mean of 0, and a standard deviation of 0.5. Let's inspect code block 32. First, lines 2–4 create two new columns that are rescaled versions of Feedback and Hours. We call these two new columns Feedback.std and Hours.std—"std" simply means that the variables have been standardized. The function rescale() comes from the arm package (Gelman and Su 2018) and does all the calculations for us.[20] The function knows how to deal with Feedback, for example, even though this variable is not numeric right now.

Lines 7–11 summarize the new means and standard deviations so you can verify that the rescaled columns indeed have mean 0 and standard deviation 0.5 (i.e., these lines of code are not necessary). Finally, lines 14–17 rerun our model—this is the exact same model that we have been discussing; the only difference here is that the variables have been rescaled. The output of the model is shown in lines 22–26 (using display()). The significance of our predictors does not change when we rescale them, so this is truly the same model discussed earlier (fit_lm4).[21] Let's reinterpret our model estimates so you can see how our interpretation is affected by rescaling our predictor variables.

INTERCEPT. The intercept is technically the same as before: the predicted score when our variables are set to zero. However, zero now has a different meaning: it represents the *mean* of our variables. Here's what you need to remember: when we rescale our variables, the intercept is the (grand) mean of our response variable—run mean(longFeedback$Score) and you will see. It's easy to understand what zero means for Hours.std, since the mean of

R code

```
# Rescale variables:
longFeedback = longFeedback %>%
  mutate(Feedback.std = arm::rescale(Feedback),
         Hours.std = arm::rescale(Hours))

# Check means and SDs for rescaled columns:
longFeedback %>%
  summarize(meanFeedback = mean(Feedback.std),
            sdFeedback = sd(Feedback.std),
            meanHours = mean(Hours.std),
            sdHours = sd(Hours.std))

# Rerun model:
fit_lm4_std = lm(Score ~ Feedback.std +
                   Hours.std +
                   Feedback.std * Hours.std,
                 data = longFeedback)

# Output:
display(fit_lm4_std)

#                        coef.est  coef.se
# (Intercept)               74.68     0.38
# Feedback.std               2.52     0.75
# Hours.std                  4.81     0.77
# Feedback.std:Hours.std -3.33     1.53

# Comparing predictions:
predict(fit_lm4_std, newdata = tibble(Hours.std = 0,
                                      Feedback.std = 0.5))
predict(fit_lm4, newdata = tibble(Hours = mean(longFeedback$Hours),
                                  Feedback = "Recast"))
```

CODE BLOCK 32 Rescaling Variables in a Linear Model

Hours is easily calculated ($\bar{x} = 10.3$). But it's harder to picture what zero means for Feedback.std—let's discuss this estimate next and then come back to the intercept.

FEEDBACK.STD. Our effect here is $\hat{\beta} = 2.52$. What does that mean? In a nutshell, a change in *1 unit* of Feedback.std increases the predicted score by 2.52 points. Another way to say this is this: the difference between explicit correction and recast, assuming that Hours is kept at its mean value, is 2.52 points. In a rescaled variable, "a unit" is represented by 2 standard deviations. Feedback.std goes from −0.5 to +0.5, and its standard deviation is 0.5. Therefore, 1 unit = $2s$ = 1.0. Simply put, a unit for Feedback.std is the "distance" between Explicit correction and Recast. The effect size in question ($\hat{\beta} = 2.52$) is what happens to Score if we go from −0.5 to +0.5—see Fig. 6.7.

To understand the effect of Feedback.std taking the intercept into consideration, we multiply the estimate in question by ±0.5, depending on which feedback group you want to consider—as is shown in Fig. 6.7. For example, if we're considering Explicit correction, we multiply $\hat{\beta} = 2.52$ by -0.5.

Fig. 6.7 illustrates how rescaling a binary variable affects the way we interpret said variable in light of the new meaning of our intercept. To sum up: the intercept is the predicted score when all other variables are set to zero. Rescaling variables means that now zero represents the mean of each variable. For a binary variable such as Feedback.std, the mean is simply the average between its levels, explicit correction and recast—hence the circle in the middle of the range in Fig. 6.7. The estimate of a rescaled binary variable represents the effect of going from one end of the range (-0.5) to the other end of the range (+ 0.5).

HOURS.STD. This estimate ($\hat{\beta} = 4.81$) represents the change that 1 unit of Hours.std will have on a predicted score. But remember that 1 unit now is no longer "1 hour". Rather, it's 2 standard deviations of Hours. The standard deviation of Hours is about 2.5, so 1 unit here represents 5 hours. Therefore, we can conclude that by studying 5 more hours, your score is predicted to increase 4.81 points (assuming that Feedback = 0, which now represents the average of both levels in the factor). The intercept assumes average values for all predictors. Therefore, it already assumes that Hours = 10.3.

FEEDBACK.STD:HOURS.STD. Like before, the interaction here is negative and tells us that the effect of Hours.std is weakened if we go from Feedback.std = –0.5 to Feedback.std = +0.5. A simple way to look at the negative estimate here ($\hat{\beta} = -3.33$) is this: if Feedback.std is held at 0 (its mean), increasing Hours.std by 1 unit ($2s \approx 5$ hours) will increase Score by 4.8 points, which is its effect size ($\hat{\beta}$)—more specifically, Score will go from 74.7 to 79.5 (you can test this using predict()). But if we assume Feedback.std = –0.5 (explicit correction), then the effect of a unit of Hours.std will be different, since the variables interact. For Hours = 0, the predicted score is given in 6.4. And for Hours = 1, the predicted score is given in 6.5. The difference here is 6.5 points, not 4.81 points. As we can see, the increase in Score as a function of

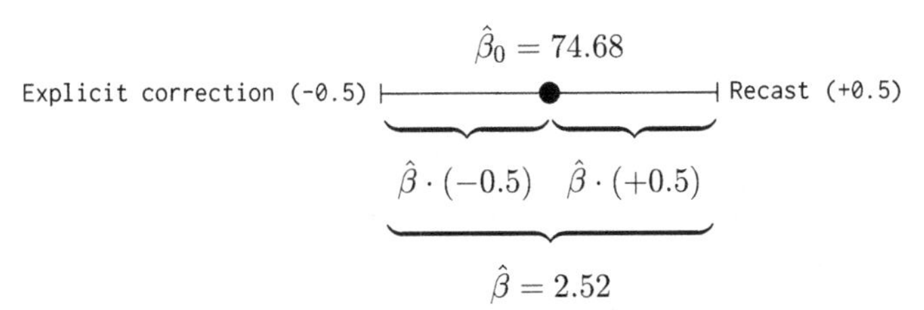

FIGURE 6.7 Rescaling a Binary Variable: Feedback.std

Hours.std depends on whether a participant is in the explicit correction group or in the recast group. This is not new, of course, since we have already discussed this interaction. What's new are the units of the estimates and how we interpret them.

$$
\begin{aligned}
\hat{y}_A &= \hat{\beta}_0 + \hat{\beta}_1 x_{\text{Feedback.std}} + \hat{\beta}_2 x_{\text{Hours.std}} + \hat{\beta}_3 x_{\text{Feedback.std}} x_{\text{Hours.std}} \\
\hat{y}_A &= 74.68 + 2.52 \cdot (-0.5) + \cancel{4.81 \cdot 0} + \cancel{(-3.33 \cdot (-0.5) \cdot 0)} \\
\hat{y}_A &= 73.4
\end{aligned}
\tag{6.4}
$$

$$
\begin{aligned}
\hat{y}_A &= \hat{\beta}_0 + \hat{\beta}_1 x_{\text{Feedback.std}} + \hat{\beta}_2 x_{\text{Hours.std}} + \hat{\beta}_3 x_{\text{Feedback.std}} x_{\text{Hours.std}} \\
\hat{y}_A &= 74.68 + 2.52 \cdot (-0.5) + 4.81 \cdot 1 + (-3.33 \cdot (-0.5) \cdot 1) \\
\hat{y}_A &= 79.9
\end{aligned}
\tag{6.5}
$$

Finally, lines 29–32 use the predict() function to illustrate the equivalence of variables and their rescaled counterparts. Asking for the predicted score using fit_lm4_std, we have Hours.std = 0 and Feedback.std = 0.5. That is equivalent to asking for the predicted score using fit_lm4 assuming Hours = 10.3 (its mean) and Feedback = Recast. Hopefully it's clear now that these two are saying exactly the same thing, only using different scales.

Now that we have rescaled our variables, which predictor is stronger: Feedback or Hours? Clearly, Hours trumps Feedback, given that its effect size is nearly twice as large as that of Feedback. A 1-unit change in Feedback.std (i.e., going from explicit correction to recast) increases the score of a participant by 2.52 points, whereas a 1-unit change in Hours.std (i.e., about 5 hours) increases the score of a participant by 4.81 points. However, bear in mind that this interpretation is ignoring the interaction between the two variables. In other words, Hours having a larger effect size than Feedback does not mean that *any* number of hours will suffice for a learner in the explicit correction group to surpass a learner in the recast group (recall the discussion we had earlier). Ultimately, to predict actual scores we need to consider the whole model, not just one of its estimates. This is yet another reason that inspecting Fig. 6.5 can help us understand interactions.

In summary, when we rescale our variables, the estimates in our model are directly comparable, since they all rely on the same unit (2 standard deviations). Once zero represents the means of our predictors, the intercept also takes a new meaning. Note, too, that you can generate Fig. 6.6 using fit_lm4_std instead of fit_lm4 in code blocks 30 and 31, so your figure for the model estimates would show rescaled estimates that are comparable. A potential disadvantage of rescaling variables is that our interpretation of the effect sizes no longer relies on the original units—for Hours.std we are no longer talking about actual hours, but rather about 2 standard deviations of the variable, so we are required to know what the standard deviation is for said variable if we want to translate 1 unit into actual hours for the reader. You could, of course, provide both sets of estimates and use the rescaled version to draw direct comparisons across estimates and use the non-rescaled estimates to interpret the effects of individual estimates, as we did with all previous models discussed in this chapter. In general, SLA studies report unscaled estimates—the vast majority of studies don't rescale variables, in part because most studies don't employ full-fledged statistical models currently.

6.4 Summary

In this chapter we discussed the basic characteristics and assumptions of linear models, and we saw different examples of such models using our hypothetical study on the effects of feedback. We also discussed how to report and present our results, how to compare different models, and how to rescale our predictor variables so that we can directly compare the magnitude of their effects. It's time to review the most important points about linear models.

- A linear regression models a continuous response variable (e.g., Score) as a function of one or multiple predictor variables, which can be categorical (e.g., Feedback) or continuous (e.g., Hours), for example. Linear models can be represented as $y_i = \beta_0 + \beta_1 x_i + \epsilon_i$, where y_i is the response value for the ith participant, β_0 is the intercept, β is the estimate of a predictor variable, x_i is the value of said variable for the ith participant, and ϵ is the error term (everything our model cannot account for). Naturally, we can have several predictor variables, so we could represent our models as $y_i = \beta_0 + \beta_1 x_{i1} + \ldots + \beta_n x_{in} + \epsilon_i$. Using matrix notation, you can also define a model as $y = X\beta + \epsilon$.
- The best line (i.e., the best model) is the one that minimizes the residual sum of squares. In other words, it's the line that minimizes the error (for any given predictor).
- Linear models assume that the relationship between our response variable and a continuous predictor variable is *linear*. They also assume that the variance of our response variable is constant across the different values of our predictor variables. This means that our model Score ~ Feedback (fit_lm2), for example, assumes that the variance of Feedback is roughly the same for the explicit correction group *and* for the recast group. This assumption is often not met.
- Another important assumption of linear models is that their residuals are normally distributed. You can easily check that for any model by using the residuals() function in R (see §6.1).
- To run a linear model, we used the lm() function (and assigned it to a variable). To inspect the output of the model, we used summary() and display() (from the arm package). The former prints a more complete (but redundant) output, which includes *p*-values. The latter prints a more

parsimonious output that provides estimates and standard errors—recall that these two are all we need to assess the effect of a given predictor. *t*-values, *p*-values, and 95% confidence intervals can all be calculated or inferred once we have estimates and standard errors. But we can also use the confint() function to calculate 95% confidence intervals.

- To compare multiple models, we use anova(), which runs an ANOVA testing whether a more complex model is actually statistically better than a simpler model at capturing the patterns observed in the data.
- Our model estimates provide effect sizes for our predictor variables, but you should not compare different estimates directly, given that variables often have different scales (e.g., Feedback *vs.* Hours). If you want to compare them, you can rescale the variables, in which case the interpretation of their estimates will have to be adjusted accordingly. We saw earlier that once variables are rescaled using rescale(), their means are 0 and their standard deviations are 0.5. In this case, 1 unit of a rescaled variable represents 2 standard deviations of the original scale of said variable.
- Presenting the results of a model can be done in different ways, but you should provide either a table or a figure. Which one is more advantageous is ultimately a matter of taste, but figures tend to be more intuitive to interpret, assuming that they look sufficiently clear.

6.5 Exercises

Ex. 6.1 Basics

1. Explain in your own words the following terms: R^2, residuals, estimates, contrast coding.
2. What's the null hypothesis for the intercept of a model?
3. What's the null hypothesis for any given estimate in a model?

Ex. 6.2 Model Estimates

1. We have been assuming Explicit correction as our reference level in Feedback—since R defines it alphabetically for us. Rerun fit_lm4, this time using Recast as the reference level for Feedback. How does the output of the model change? How about your interpretation of the estimates? *Hint:* Consult §6.2.2.
2. Generate Fig. 6.6 for fit_lm4_std using the sjPlot package—all you need is at the bottom of code block 31. Report the results.

Ex. 6.3 Predicting New Data

1. Using the predict() function on fit_lm4, generate predicted scores for participants who study 8, 10, 15, and 25 hours per week (consider both

feedback groups). How do they differ? *Hint:* Predicted scores are not affected by your choice of reference level for Feedback.

2. How do the predicted scores help us see the interaction between Hours and Feedback?

Ex. 6.4 More Variables

1. Run a new model that adds L1 as a predictor (on top of the predictors in fit_lm4). Call it fit_lm5. Interpret the model, and compare its fit to fit_lm4—notice the reference level for L1 (which you may or may not want to change).
2. Does adding L1 improve the model? Are you surprised, given the estimates for L1?

Notes

1. You could, of course, source continuousDataPlots.R instead, which is the script that contains code block 11. That script, however, is in a different folder/working directory, which means you would need the complete path to it.
2. Like the true mean of a population (μ), we don't know the true value of ϵ, but we can estimate it ($\hat{e}$).
3. Don't try to run this code just yet. First, we need to run a model and assign it to a variable. Then, you should replace MODEL in the code with said variable.
4. Naturally, we can have multiple variables in a model, in which case we will have multiple $\hat{\beta}$s. We will discuss such examples later.
5. Technically, what we're running is lm(Score ~ 1 + Hours, data = longFeedback), where 1 represents the intercept. R simply assumes the intercept is there, so we can leave 1 out.
6. t-values give us the magnitude of a difference in units of standard error, which means t- and p-values are connected to each other.
7. I will often propose that we manually calculate confidence intervals in this book as an exercise to review SEs and CIs. However, you should simply use confint() instead.
8. In general, we should report the adjusted R^2, which takes into account the number of predictors that a model has. This will be especially important later when we add multiple predictors to our models.
9. Remember to install and load the arm package. You should already have this package if you added code block 25 to linearModels.R.
10. As already mentioned, R will order the levels alphabetically and choose the first one as the intercept. We can change that by running longFeedback = longFeedback%>%mutate(Feedback = relevel(as.factor(Feedback), ref = "Recast")). Then rerun the model and check the output again—note that here we first make Feedback a factor, and then we change its reference level.
11. You could simply report $\hat{\beta} = 2.9, SE = 0.77$, but this may be too minimalistic for journals in our field.
12. Note that it makes sense to use facets for our categorical variable and leave the x-axis for our continuous variable.
13. By adding more variables to a model, our R^2 is going to increase. That's why we should consider the adjusted R^2, which has been adjusted for the number of predictors in the model.

14. How do you think the different R^2 values from our two previous models could help us guess the relative importance of the predictors in question?
15. We could use shape or color, for example, to make the levels of Feedback look different, but it's not easy to visually see that difference given the number of data points we have. Clearly, the focus of Fig. 6.5 is to show the different trend lines.
16. If you use LaTeX to produce your academic documents, you can use packages such as xtable and memisc in R to generate tables ready for publication.
17. Notice that this is what we typically report for ANOVAs. That makes sense, given that we are comparing variances across models. And much like the ANOVAs you may be familiar with, we could compare multiple models at once, not just two.
18. Unlike the previous code blocks, which you should place inside linearModels.R, you should place these two code blocks inside plotsLinearModels.R, so as to keep plots and models in separate scripts.
19. Some people scale variables by dividing by *1* standard deviation. To understand why we will divide by *2* standard deviations instead, see Gelman (2008b).
20. See Appendix A on the use of "::" in functions. Here, the function rescale() is present in the arm package and also in the scales package (which we use for adding percentages to our axes, among other things).
21. Note that the significance of the intercept may change, given that its meaning is different now.

7

LOGISTIC REGRESSION

All the models we have explored thus far involve a *continuous* response variable (Score in chapter 6). But in second language research, and linguistics more generally, we are often interested in binary response variables. These variables are essentially 0/1, but we will usually label them incorrect/correct, no/yes, non-target-like/target-like. If you recall our data on relative clauses in chapter 4 (rClauseData.csv), the Response variable was also binary: Low or High (and we had some NAs as well). Clearly, this is a very different situation when compared to continuous variables: while scores can have multiple values along a continuum (e.g., 78.4), binary variables cannot—by definition.

In this chapter we will learn how to analyze binary response variables by running *logistic regressions*—see Jaeger (2008) on why you should not use ANOVAs for categorical response variables. Logistic regressions are underlyingly very similar to linear regressions, so a lot of our discussion will be familiar from chapter 6. However, there are important differences, given the different nature of our response variable. As with chapter 6, we will first go over some basics (§7.1) and then we'll see how to run and interpret logistic models in R (§7.2). You may want to go back to §7.1 after reading §7.2 to consolidate your understanding of logistic models—you shouldn't be surprised if you need to reread about the same statistical concept several times!

Before we get started, let's create some scripts. You should place these scripts in the Frequentist folder we created in chapter 6, so all the files associated with our models will be in the same directory. You should have rClauseData.csv in there too, since we'll be running our models on the dataset used in chapters 4 and 5. We will create three scripts: dataPrepCatModels.R, plotsCatModels.R, and logModels.R—we already have our R Project, Frequentist.RProj, in the same directory, so we can use a single R Project to manage all the files connected to the Frequentist statistical models in this book. These scripts follow the same rationale we applied to chapter 6: one script will prepare the data, another will plot the data, and the third script will run the models. Could you do all three tasks in a single script? Absolutely. Should you? Probably not. As our analyses become more complex, our scripts will get substantially longer. Separating our different tasks into shorter scripts is a healthy habit.

DATA FILE

In this chapter (§7.2) we will use rClauseData.csv again. This file simulates a hypothetical study on relative clauses in second language English.

7.1 Introduction

Observe Fig. 7.1, which plots participants' reaction time (*x*-axis) against their native language (L1, on the *y*-axis)—these data come from rClauseData.csv. Recall that in the hypothetical study in question, Spanish speakers are second language learners of English. Here our response variable is binary (L1), and our predictor variable is continuous (reaction times)—so this is the opposite of what we had in chapter 6. The intuition is simple: second language learners are expected to take longer to process sentences in the L2, so we should see slower reaction times for learners relative to native speakers—see, e.g., Herschensohn (2013) for a review of age-related effects in second language acquisition. That's what we observe in the box plot: Spanish speakers (L2ers) have longer reaction times than English speakers (native group).

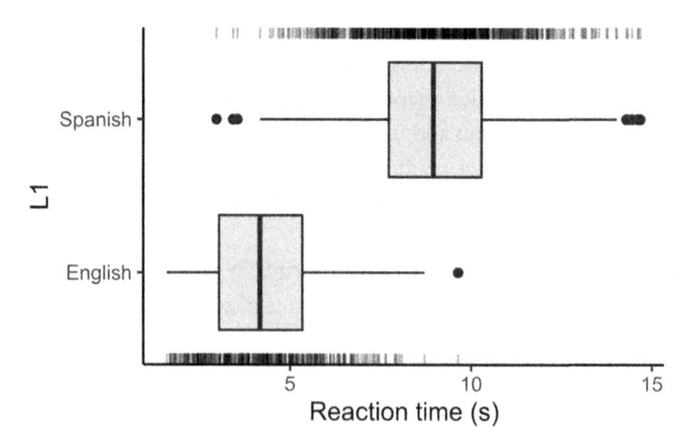

FIGURE 7.1 Native Language As a Function of Reaction Time: L1 ~ RT

The first important characteristic of binary response variables is that we want to predict *probabilities*. In other words, given a specific reaction time from participant *i*, what's the probability that *i* is a speaker of Spanish (L2er)? You may be wondering whether it would be possible to analyze this data with a linear model using the lm() function in R. After all, probabilities are numbers, and if we're predicting numbers, we could simply use a linear model. Let's entertain that possibility for a second.

Examine the left plot in Fig. 7.2. The *y*-axis no longer says "Spanish" and "English". Instead, we treat "Spanish" as 1 and "English" as 0. Here we're using geom_point() to plot our data, but because the *y*-axis (our response variable) is binary, all the data points are clustered either at 0 or at 1. This is not a very informative plot in general—recall our discussion on plotting categorical data in chapter 4, where we transformed our data before plotting.

Running a linear model means fitting a straight line to the data, and that's what we see in the left plot in Fig. 7.2. Do you see any problems? Probabilities can only go from 0 to 1, but our straight line ignores that completely. As a result, if a participant has a reaction time of 13 seconds, the probability that this participant is a second language learner is predicted to be about 1.36—which makes no sense. What we actually want is the curve in the right plot in Fig. 7.2. The S-shaped line in question, also known as a *logistic curve*, guarantees that our predicted values will be between 0 and 1.

The take-home message here is simple: a linear regression fits a straight line to our data (left plot in Fig. 7.2), while a logistic regression fits a *curved* line to our data (right plot in Fig. 7.2). These different approaches result from the fact that a linear regression models a continuous response variable, while a logistic variable predicts the *probability* that a binary variable will be 1, which here represents a learner of English (native speaker of Spanish).[1]

At this point, you are already familiar with a model that fits a straight line to the data (chapter 6): $y = \beta_0 + \beta x + \epsilon$. Now, instead of predicting a continuous variable, we want to estimate a probability *P*. The question is how to use what we already have (our linear model) to predict a probability, that is, to generate an S-shaped curve. The trick here is to use the *logit function*: $logit(P) = \beta_0 + \beta x$.

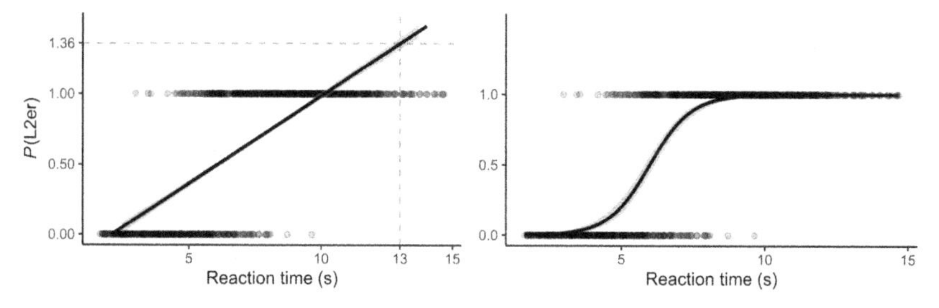

FIGURE 7.2 Modeling a Binary Response As a Function of a Continuous Predictor

The logit of a probability is equivalent to the log of the odds of said probability. For example, a 50% probability is equivalent to 1/1 odds; an 80% probability is equivalent to 4/1 odds. The generalization is $odds = \frac{P}{1-P}$. In other words, the odds of *X* happening equals the probability of *X* happening divided by 1 minus that probability. Thus, if $P = 0.8$, $odds = \frac{0.8}{1-0.8} = 4$. The logit function simply takes the natural log ($log_e x$) of the odds, represented as $ln()$, that is, $ln(odds) = ln\left(\frac{P}{1-P}\right)$. Table 7.1 maps some probabilities to odds to log-odds, that is, ln(odds).

But why do we need log-odds? First, let's quickly review the notion of odds. Odds are simply the ratio of something happening (e.g., being a learner) to something not happening (e.g., not being a learner). So if the odds of being a learner given a specific reaction time are 3 to 2, that gives us $odds = \frac{3}{2} = 1.5$. Probabilities, on the other hand, are the ratio of something happening (e.g., being a learner) to everything that could happen (e.g., being a learner *and* not being a learner). Thus, the probability of being a learner is $P = \frac{3}{3+2} = 0.6$—this comparison is shown in Table 7.1. Why then do we take the *log* of the odds?

Let's suppose we're discussing the odds of *not* being a learner. Here, the odds go from 0 to 1. If, on the other hand, we're talking about the odds of being a learner, our range goes instead from 1 to ∞. In other words, odds have asymmetrical magnitudes. For instance, if the odds are against being a learner 1 to 4, that's 0.25; if they are in favor 4 to 1, that's 4. Intuitively, these should be symmetrical—but on the odds scale they're not. Once we take the log of the odds, we make our scale symmetrical. For example, $ln(0.25) = -0.60$, and $ln(4) = +0.60$. This is shown in Fig. 7.3—bullets illustrate the example just discussed.

Now let's look back at the model, shown in 7.1. As we just discussed, log-odds are linear. In other words, unlike probabilities, which go from 0 to 1 and are not constant, log-odds go from $-\infty$ to $+\infty$, are symmetrical, and follow a straight line. Crucially, because log-odds are a straight line, we can use the same type of model we used in chapter 6. Furthermore, it's very easy to go

TABLE 7.1 Probability (*P*), Odds, and ln(odds)

P	*Odds*	*ln(odds)*
0.10	0.11	−2.20
0.20	0.25	−1.39
0.30	0.43	−0.85
0.40	0.67	−0.41
0.50	1.00	0.00
0.60	1.50	0.41
0.70	2.33	0.85
0.80	4.00	1.39
0.90	9.00	2.20

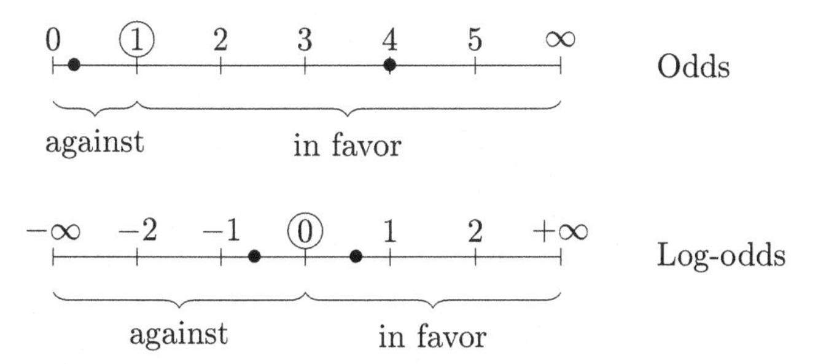

FIGURE 7.3 Odds and Log-odds

from log-odds to probabilities—and vice versa (as we can see in Table 7.1). We are essentially transforming our binary variable into a continuous variable (log-odds) using the logit function, so that later we can go from log-odds to probabilities (our curve in Fig. 7.2). Ultimately, it's much easier to work with a straight line than with an S-shaped curve (also known as sigmoid curve). In summary, while a linear regression assumes that our predicted response is a linear function of the coefficients, a logistic regression assumes that the predicted *log-odds* of the response is a linear function of the coefficients.

$$\begin{aligned} ln(odds) &= ln\left(\frac{P}{1-P}\right) = logit(P) \\ logit(P) &= \beta_0 + \beta_1 x_{i1} + \ldots + \beta_n x_{in} \\ P &= logit^{-1}(\beta_0 + \beta_1 x_{i1} + \ldots + \beta_n x_{in}) \\ P &= logit^{-1}(X_i \beta) \end{aligned} \tag{7.1}$$

$$odds = e^{ln(odds)} \tag{7.2}$$

$$P = \frac{e^{ln(odds)}}{1 + e^{ln(odds)}} \tag{7.3}$$

If you look closely at 7.1 you will notice that logistic regressions will give us estimates in ln(odds), that is, *logit*(*P*). To get odds from log-odds, we exponentiate the log-odds—see 7.2. To get probabilities from log-odds, we use the inverse logit function ($logit^{-1}$), which is shown in 7.3. That's why understanding log-odds is crucial if you want to master logistic regressions: when we run our models, $\hat{\beta}$ values will all be in log-odds—we'll practice these transformations several times in §7.2. Don't worry: while it takes some time to get used to ln(odds), you can start by using Table 7.1 as a reference. If you have positive log-odds, that means the probability is above 50%. If you have negative log-odds, the probability is below 50%. We will see this in action in §7.2.

Hopefully it's clear at this point why we want to use log-odds instead of probabilities. Although probabilities are more intuitive and easier to understand, log-odds are linear. If you look once more at the right plot in Fig. 7.2, you will notice that how much the y-axis changes in an S-shaped curve depends on *where* you are on the x-axis. For example, the change in $P(L2er = 1)$ from 1 to 2 seconds on the x-axis is nearly zero, but the change is huge if you go from 5 to 6 seconds (the curve is steepest right around 5–8 seconds). So while probabilities are indeed more intuitive, they are also dependent on where we are on a curve, and that's not ideal since our answer will change depending on the value on the x-axis. Log-odds, on the other hand, offer us an estimate that is constant across all values on the x-axis.

7.1.1 Defining the Best Curve in a Logistic Model

You may have noticed that we don't have ϵ in 7.1. When we discussed linear regressions in chapter 6, we spent some time covering residuals (sum of squares). We saw that we can use residuals to calculate R^2, which allowed us to see how much of variation in the data could be accounted for by the predictor we used in our model. So where are they in our logistic regression?

In logistic regressions, we don't use least squares, so the concept of residuals doesn't apply. Think about it this way: when we transform our raw data (0s and 1s) into log-odds, we push the data to $-\infty$ and $+\infty$. Consequently, the distance from our line (in log-odds) to the actual data points will be, well, infinite (!). And this doesn't really help us assess different fits, because they will all result in infinite residuals. For that reason, we won't be talking about residuals (or R^2) in this chapter—I return to this point in the summary provided at the end of the chapter.

So how do we determine the best fit in a logistic model? We use *maximum likelihood*. The details of how maximum likelihood works are not important for our purposes here. What is important is that our models will give us a fit that maximizes the probability of obtaining the empirical data being modeled. In other words, logistic models will give us the best possible fit for the data we are modeling.

7.1.2 A Family of Models

In the next section, you will notice many similarities between logistic models and the linear models we ran in chapter 6. Given our discussion earlier, that shouldn't be surprising. Both logistic and linear models are part of the same "family" of *generalized linear models*, or GLMs. In fact, the function we will use in this chapter, `glm()`, can also be used to run linear models: the models we ran in chapter 6, lm(y ~ x), could be rerun as glm(y ~ x, family = "gaussian"). The family argument here simply specifies the type of generalized linear model ("gaussian" indicates that our response variable is continuous and follows a normal distribution). For logistic regressions, we will specify family = "binomial", since our response variable follows a binomial distribution.[2]

7.2 Examples and Interpretation

Before we run our first model, let's prepare our data. Add the lines in code block 33 to dataPrepCatModels.R. Here we will simply make some simple adjustments to our data (rClauseData.csv). First, we will remove fillers from the data, since we don't care about them (lines 8–10). Second, we will create two new columns using the mutate() function (lines 13–15). The first column, Low, will be 1 for all the low attachment responses and 0 otherwise (high attachment).[3] The second column, L2er, does the same thing for L1. These two variables (Response and L1) will be our response variables in the examples that follow, so it's crucial that we understand what exactly we're modeling. For example, if our model had L1 as its response variable, are we predicting the probability that L1 = English or that L1 = Spanish?

Recall that R will order the levels of a factor alphabetically. Thus, if we run our models using the Response and L1 columns, R will automatically assign 0 to Response = High and 1 to Response = Low. Likewise, it will assign 0 to L1 = English and 1 to Spanish. It turns out that this is exactly what we want, since we want to focus on the learners here. But we won't always be lucky. Lines 13–15 show you how to create new binary columns where *you* decide which level of a factor will be 1 or 0. We do that by using the extremely useful function ifelse(), which requires three arguments: ifelse(condition, TRUE, FALSE). First, we add a condition (Response == "Low") in line 14. Second, we define what value we want if the condition is TRUE—here, we want 1. Third, we define what value we want if the condition is FALSE—we do the same in line 15 for the L2er column. We can ask R to return whatever we want—and here we just want 0/1. In plain English, we're saying "if the response is Low, insert 1; otherwise, insert 0."

Lines 18–31 in code block 33 will be useful for chapter 8, so we won't get into them now. Finally, line 34 simply transforms all the columns that belong to the chr (character) class into factors using the mutate_if() function—recall that you can check variable classes with str().

In the examples that follow, we will visualize our data first and then model it—as usual. You should place code blocks that generate figures in plotsCatModels.R and code blocks that run models in logModels.R, so we can keep models and figures separated.

7.2.1 Can Reaction Time Differentiate Learners and Native Speakers?

Our first model will be based on Fig. 7.1, where we plotted L1 on the *y*-axis and reaction times on the *x*-axis. The figure clearly showed that native speakers and L2ers have different reaction times overall. What we want to do here is assess the effect size of reaction time (alternatively, we want to model the probability that participant *i* is a learner as a function of *i*'s reaction time): $P(L2er = 1) = logit^{-1}(\beta_0 + \beta_1 \cdot RT_i)$. Let's spend some time examining code block 34.

First, notice that line 1 sources dataPrepCatModels.R—see code block 33. Line 3 loads the arm package (the same package we used in chapter 6 to print a more minimalistic output: display()). Next, in lines 5 and 6, we run our model and print its output using summary(). As you can see, the code is practically identical to what we did in chapter 6 for our linear models.

R code

```
1  # Remember to add this code block to dataPrepCatModels.R
2  library(tidyverse)
3
4  # Import our data:
5  rc = read_csv("rClauseData.csv")
6
7  # Remove fillers:
8  rc = rc %>%
9    filter(Type != "Filler") %>%
10   droplevels()
11
12 # Create 0/1 column for response and for L1:
13 rc = rc %>%
14   mutate(Low = ifelse(Response == "Low", 1, 0),
15          L2er = ifelse(L1 == "Spanish", 1, 0))
16
17 # Make certainty ordered factor (for next chapter):
18 rc = rc %>%
19   mutate(Certainty = factor(Certainty, ordered = TRUE))
20
21 # Mutate certainty into three categories: 1-2, 3-4, 5-6:
22 rc = rc %>%
23   mutate(Certainty3 = ifelse(Certainty < 3, "Not certain",
24                              ifelse(Certainty > 4, "Certain",
25                                     "Neutral")))
26
27 # Adjust order of levels:
28 rc = rc %>%
29   mutate(Certainty3 = factor(Certainty3,
30                              levels = c("Not certain", "Neutral", "Certain"),
31                              ordered = TRUE))
32
33 # Make character columns factors:
34 rc = rc %>% mutate_if(is.character, as.factor)
```

CODE BLOCK 33 Preparing our Data for Logistic Models

R code

```
source("dataPrepCatModels.R")

library(arm) # to use invlogit function in line 25

fit_glm1 = glm(L2er ~ RT, data = rc, family = "binomial")
summary(fit_glm1)

# Output excerpt using summary()
# Coefficients:
#              Estimate Std. Error z value Pr(>|z|)
# (Intercept)   -8.270     0.618  -13.38   <2e-16 ***
#   RT           1.381     0.098   14.09   <2e-16 ***
#   ---
#   Signif. codes:  0 '***' 0.001 '**' 0.01 '*' 0.05 '.' 0.1 ' ' 1
#
# Null deviance: 1145.73  on 899  degrees of freedom
# Residual deviance:  398.01  on 898  degrees of freedom
# AIC: 402.01

# From log-odds to odds:
exp(coef(fit_glm1)[["(Intercept)"]])
exp(coef(fit_glm1)[["RT"]])

# From log-odds to probabilities (interpret with caution!):
invlogit(coef(fit_glm1)[["(Intercept)"]])

# Probability change: from 1s to 10s:
predict(fit_glm1,
        newdata = tibble(RT = seq(from = 1, to = 10)), type = "response")
```

CODE BLOCK 34 Running a Logistic Regression in R

Let's now interpret our estimates in lines 11–12. First, we can see that both our intercept and reaction time (RT) (i.e., both $\hat{\beta}_0$ and $\hat{\beta}$) are significant ($p <$ 0.001). What does that mean? The intercept tells us the predicted log-odds of being a learner assuming a reaction time of 0 seconds. This should sound familiar from chapter 6—except for the log-odds bit. Intercepts are always the predicted response when all other variables are 0. The estimate for RT, in turn, tells us how much the log-odds of being a learner change as we increase 1 unit of RT (i.e., 1 second).

The very first thing you should note (besides the significance) for each estimate is the sign: the intercept is *negative* here, while RT is *positive.* If the intercept is negative, it basically means that the probability of being a learner if your reaction time is 0 is lower than 50%. If RT is positive, it means that as you increase reaction time, the probability of being a learner also increases. The nice thing about focusing on the sign is that you don't need to think about log-odds to understand the overall pattern here: reaction time is positively correlated with the probability of being a learner, which makes sense given Fig. 7.1, where learners had slower reaction times than native speakers.

Next, let's focus on the actual effect sizes. If you look back at Table 7.1, you will notice that we go from −2.20 to 2.20 (log-odds). That range covers probabilities between 0.10 and 0.90. Now look at the estimate for our intercept: it's less than −8 (!). That's a *very* small number: if −2.20 log-odds is equivalent to a probability of 0.10, we already now the probability of being a learner if your reaction time is 0 will be tiny—if you run line 25 you will find out how tiny it is. You can also manually calculate that probability using 7.3: $p = \frac{e^{-8.27}}{1+e^{-8.27}} = 0.000256$. This is equivalent to running `exp(−8.27)/(1 + exp(−8.27))` or `invlogit(-8.27)` in R. The specific estimate for the intercept here is not very meaningful, because we don't expect a participant to have a reaction time of 0 seconds. Let's move on to the effect of RT, our focus here.

The estimate for RT is $\hat{\beta} = 1.38$ (log-odds). Let's first interpret this in terms of odds by taking the exponential of the estimate ($e^{|\hat{\beta}|}$), or `exp(1.38)` in R—line 22 of code block 34 calculates that for us (line 21 does the same for the intercept). An increase (positive sign) of 1.38 log-odds is equivalent to an increase by a factor of 3.97. In other words, as you increase reaction time by 1 unit (1 second), the odds of being a learner go up by a factor of almost 4—which is a lot!

How about the change in probability? Here, the answer can be a little tricky. Remember that a probability curve is *not* a straight line, which means the change in probability is not constant across all values of the predictor variable—we discussed this earlier in reference to the right plot in Fig. 7.2. To make things more concrete, then, let's pick actual reaction times to see how much the probability of being a learner changes as a function of our predictor variable here.

Take a look at lines 28–29 in code block 34. The function in question, `predict()`, should be familiar from chapter 6: we can use it to predict new data given a model fit. This time, we're using an additional argument, namely, `type = "response"`—which will give us probabilities (as opposed to log-odds). Here, our new data is a sequence of reaction times: from 1 to 10 seconds, with 1-second intervals (i.e., 1, 2, ..., 9, 10). What we're doing is simple: given the model we just ran, what's the probability that a participant is a learner if his or her reaction time is 1 second? We ask that for all ten reaction times in question.

Run lines 28–29. You will see that, for a reaction time of 5 seconds, the probability of being a learner is 20%. For 6 seconds, it's 50% ($P(L2er = 1) = 0.50$). Notice that this change in 1 second caused a positive difference of 30%. Now look at 9 and 10 seconds, which result in 0.98 and 0.99. Here, the change in probability is minuscule in comparison to the change between

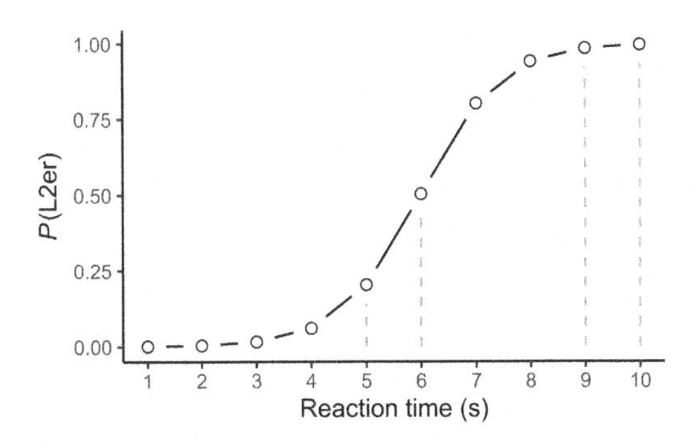

FIGURE 7.4 Predicted Probabilities for Ten Reaction Times (Using Model Fit)

5 and 6 seconds. This makes sense: our probabilities do *not* follow a linear trend, so the biggest change in probability will occur in the middle of our S-shaped curve. Fig. 7.4 plots the predicted probabilities from lines 28–29. That's why simply taking the probability of the effect size can be misleading: it all depends on *where* we look along the *x*-axis.

Let's now examine the remainder of the output given in code block 34. Lines 16 and 17 tell us the null deviance and the residual deviance of the model, respectively. The null deviance basically tells us how accurate our predictions would be if we only had the intercept in the model (i.e., no predictor variable at all). The residual deviance then tells us what happens once we include our predictor variable, RT. Here, the deviance goes from 1145 to 398—this is a substantial difference, which in turn indicates that reaction time indeed helps our model's accuracy at predicting the probability that a given participant is a learner, that is, $P(L2er = 1)$.

Finally, in line 18, our output also gives us the AIC, or *Akaike information criterion* (Akaike 1974). Here, AIC = 402, which in and of itself doesn't tell us much (unlike R^2 in linear models). However, this number will be useful once we start comparing models: the lower the AIC of a model, the better the fit. So if we run three different models on the same data, the one with the lowest AIC will have the best fit of the three.

AIC helps us select models by estimating the loss of information involved in the process of fitting a model. Think of a model as a representation of our data—a representation that is never perfect. We want our models to be maximally good, but at the same time we want them to be as simple as possible to avoid overfitting the data. Here's why: a model with too many variables will fit the data better than a model with fewer variables, but in doing that our model may be picking up noise that is specific to our data and that will not be present in a different sample of participants, for example. If we design a model that works perfectly for the data we have, it may fail to work properly on data we don't have yet, but which may be collected later. Ideally, we want our models to capture the patterns in our data but also to predict future patterns. Consequently, what we want is a compromise between a model that is too good (and which will therefore overfit the data) and a model that is too simple (and which will therefore *under*fit the data). AIC values help us decide which model offers the best compromise. We will return to this discussion shortly.

One additional way to see how good or accurate our fit is would be to compare its predictions to the actual data. In other words, we could compare the value of L2er in rc to what the model predicts for each reaction time in the data. That's what we'll do next.

Code block 35, which you should add to logModels.R, calculates the accuracy of our model, fit_glm1. Lines 2–9 create a new tibble, fit_glm1_accuracy. First, we take rc (line 2), and select the only two variables that we care about right now (line 3), namely, L2er and RT. Next, in lines 4–9, we create three different columns. Column 1, called pL2er, will contain the predictions (in probabilities) of fit_glm1 for all the reaction times in the data (line 5). Line 7 then dichotomizes these predictions in the form of a new column, L2erBin: every time a predicted probability is above 0.5, we classify it as 1 (meaning: if your probability of being a learner is above 50%, you're a learner). When the probability is under 50%, we classify L2erBin as 0. And for situations where the predicted probability is exactly 50%, we'll have NAs.

After we have dichotomized the predictions of our model, our next task is simple: we create a new column, Accuracy, and for every time our dichotomized prediction matches the actual value of L2er, Accuracy = 1 (that is,

R code

```
1  # Measuring accuracy of fit_glm1:
2  fit_glm1_accuracy = rc %>%
3    dplyr::select(L2er, RT) %>%
4    mutate(pL2er = predict(fit_glm1,
5                           newdata = tibble(RT = RT),
6                           type = "response"),
7           L2erBin = ifelse(pL2er > 0.5, 1,
8                            ifelse(pL2er < 0.5, 0, NA)),
9           Accuracy = ifelse(L2erBin == L2er, 1, 0))
10
11 fit_glm1_accuracy %>%
12   summarize(Correct = sum(Accuracy)/nrow(.))
13
14 # A tibble: 1 x 1
15 #     Correct
16 #       <dbl>
17 #   1   0.908
```

CODE BLOCK 35 Assessing Model's Accuracy: fit_glm1

our model is correct for that prediction). Finally, in lines 11–12, we count how many times we have 1s in our Accuracy column and divide that number by the number of rows in the data (nrow(.)). Line 17 has the answer: our model's predictions were correct more than 90% of the time, which is impressive considering that we only have a single predictor variable.

There are two reasons that we shouldn't be too impressed with our accuracy here. First, the data being modeled is hypothetical. Your actual data may look very different depending on a number of factors (who your participants are, how you designed your experiment, etc.). Second, we used all of our data to run a model and then gave it *the same* data to measure its accuracy. This is like giving students a test today for them to study for a test next week and then giving them the exact same test next week (!). No one would be surprised if they did well.

Is it a problem that we're testing our model's accuracy with the same data we used to run the model? It depends. In machine learning, for example, we train the model with one dataset and then test it on a *different* dataset. In that case, yes, this would be a problem. But not all statistical analyses have machine learning in mind, so you may not care that your model is trained on the same data with which you're testing it. If your intention is to have a percentage to say how much of your data your model accounts for, then calculating its accuracy like we did is a perfectly reasonable option to entertain.

REPORTING RESULTS

A logistic regression confirms that reaction time is a significant predictor of whether or not a participant is a learner (L2er): $\hat{\beta} = 1.38, p < 0.0001$. The estimate indicates that for every additional second of a participant's reaction time, his/her odds of being an L2er go up by a factor of 3.97 ($e^{|1.38|}$). The model in question accurately predicts whether or not a participant is an L2er 90% of the time in the data modeled.

Recall that you can also report the standard error of the estimates or the 95% confidence intervals of the estimates. If you use the display() function to print the output of fit_glm1, you will notice again that only estimates and standard errors are provided, given that these two numbers are sufficient to calculate (or infer) the z-value,[4] the p-value, and the 95% confidence intervals.

Now that we have examined our first logistic regression in detail, we won't need to repeat all the steps from fit_glm1 when discussing our next models. For example, we won't calculate the accuracy of the model—but you can easily adapt the code in code block 35. Remember: you can always come back to our first model if you feel like reviewing some of the details involved in interpreting a model's estimates. But don't worry: we will run and interpret three more models in this chapter.

7.2.2 *Does* Condition *Affect Responses?*

Recall that the hypothetical study in rc involves relative clauses in English whose heads have an ambiguous interpretation. While Spanish is known to favor *high* attachment, English is known to favor *low* attachment (chapter 4). Let's take another look at the sentence *Mary saw the daughter*$_{\text{NP1 = high}}$ *of the nurse*$_{\text{NP2 = low}}$ *who likes to dance*, first presented in (1) back in chapter 4. Here, a Spanish speaker would likely assume that *who* refers to *the daughter*, while an English speaker would likely assume that it refers to *the nurse*.

Our main question here, of course, is whether a pause after the two NPs in question (*the daughter* and *the nurse*) could affect speakers' preferences. We can also ask questions about proficiency effects, or L1 effects (do native speakers and learners behave differently?), but here we will focus solely on the role of Condition, which is a factor with three levels: NoBreak, High, and Low—as shown in (2) in chapter 4. Later in this chapter, we will analyze other variables in more complex models. For now, let's focus on Fig. 7.5,[5] which plots the percentage of low attachment responses for all three conditions in the hypothetical study in question.

We can clearly see that different conditions yield different response patterns in the data—the dashed horizontal line demarcates the 50% mark on the *y*-axis. In the Low condition, where a break is inserted immediately after NP_2 in the stimuli, participants choose NP_2 as the head of the relative clause more than 60% of the time—i.e., most of the time. In conditions High and NoBreak, on the other hand, the preference for low attachment is under 50%.[6]

Before we actually run the model, let's think about what we expect to see in our model using what we discussed in the previous section (§7.1). First, let's take NoBreak as our reference so that we can compare High and Low to a

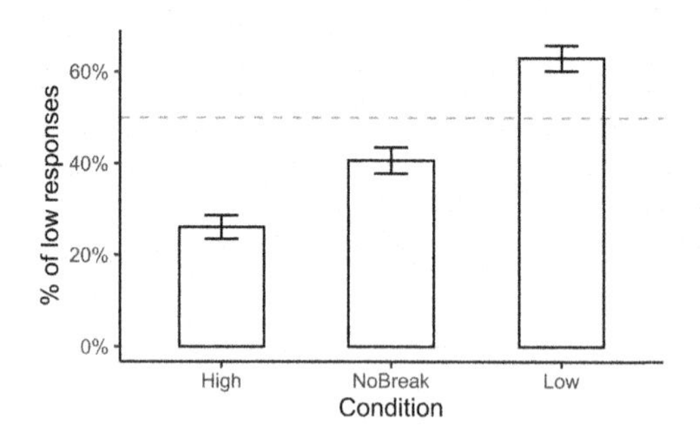

FIGURE 7.5 Preference for Low Attachment by Condition

baseline (no break at all). Look at Fig. 7.5 again: the NoBreak condition is *below* the 50% mark. If you recall Table 7.1, a 50% probability is equivalent to 0 log-odds. Because NoBreak is *below* 50%, we expect a *negative* effect size (β) for this particular condition. And because we're using NoBreak as our reference level, it will be our intercept. As a result, we predict our intercept will be negative: $\hat{\beta}_0 < 0$. Finally, given the error bars in our figure, it shouldn't surprise us if the negative effect in question is significant ($H_0 : \hat{\beta}_0 = 0$).

$$P(Low = 1) = logit^{-1}(\hat{\beta}_0 + \hat{\beta}_{high} x_{i_{high}} + \hat{\beta}_{low} x_{i_{low}}) \tag{7.4}$$

Our model can be represented in R as Low ~ Condition. Mathematically, it's represented in 7.4. If you recall our discussion in chapter 6, $x_{i_{high}}$ represents the condition of item (stimulus) i (either 1 or 0). An item can't be in two conditions at once, of course, so when an item is in one condition, we just "turn off" the terms in our model that refer to the other condition. When Condition = High, we end up with $\hat{\beta}_0 + \hat{\beta}_{high} \cdot 1 + \cancel{\hat{\beta}_{low} \cdot 0}$. And when Condition = Low, we have $\hat{\beta}_0 + \cancel{\hat{\beta}_{high} \cdot 0} + \hat{\beta}_{low} \cdot 1$. Finally, when Condition = NoBreak, both $\hat{\beta}_{high}$ and $\hat{\beta}_{low}$ will be zero, so we're left with our intercept: $\hat{\beta}_0 + \cancel{\hat{\beta}_{high} \cdot 0} + \cancel{\hat{\beta}_{low} \cdot 0}$.

If we have a factor with n levels, we will have $n-1$ $\hat{\beta}$s in our model (plus our intercept, $\hat{\beta}_0$). The intercept represents one of the levels in a factor—notice that this is not exclusive to logistic models, since we had the exact same situation back in chapter 6. In a continuous factor, the intercept represents zero; in a categorical factor, the intercept represents whichever level is set to zero (our reference). Recall that by default R will order the levels alphabetically and pick the first one as the intercept. Here, that means Condition = High, which is not ideal—in the earlier examples we assumed that NoBreak was our reference level.

Observe lines 2–3 in code block 36 (you should add this code block to logModels.R). Here we are using the relevel() function to set NoBreak as our reference level. Lines 8–20 show the output of our model, fit_glm2. Now let's interpret these results relative to Fig. 7.5.

Unsurprisingly, the intercept is negative (and significant)—so the NoBreak condition lowers the probability of choosing low attachment. To be more specific, this simply means that when no break is present, the probability of a participant choosing low attachment is below 50%—exactly what we predicted earlier given the figure.

Now let's look at the effect sizes in more detail. The effect size of our intercept is $\hat{\beta}_0 = -0.38$ (in log-odds). If we take the inverse logit of that number using the invlogit() function in R (arm package), we'll find that invlogit (−0.38) = 0.4. This is the probability of choosing low attachment in the

NoBreak condition—this number makes perfect sense if you look back at Fig. 7.5.

We also see in code block 36 that a high break has a significant effect on the probability of choosing low attachment ($\hat{\beta} = -0.67, p < 0.001$). We know the interpretation of the intercept is relative to 50%, but how about the interpretation of our other β values? These are always interpreted relative to the intercept—(see chapter 6). Look back at Fig. 7.5 and focus on the NoBreak bar. Relative to it, the bar for High is *even lower*. Which means that, relative to β_0, β_{high} should be negative, because it lowers the probability of choosing low attachment even more when compared to the NoBreak condition. To be more specific, having a high break lowers the odds of choosing low attachment by a factor of 1.95 ($e^{|\beta|} = e^{|0.67|} = 1.95$)—relative to not having a break at all.

Finally, the effect of ConditionLow is also significant ($\hat{\beta} = 0.91, p < 0.0001$). This means that the odds of choosing low attachment increase by a factor of almost 2.5 if we had a low break in the stimulus—again, relative to NoBreak. We can certainly use the predict() function here to play around with some possible values. For example, we can predict the probability of choosing low attachment in all three possible scenarios, namely, Condition = High, Condition = NoBreak, and Condition = Low. We can do that by running predict(fit_glm2, newdata = tibble(Condition = c("High", "NoBreak", "Low")), type = "response"). The answer will be 0.26, 0.41,

R code

```
# Set NoBreak as our reference for Condition:
rc = rc %>%
  mutate(Condition = relevel(as.factor(Condition), ref = "NoBreak"))

fit_glm2 = glm(Low ~ Condition, data = rc, family = "binomial")
summary(fit_glm2)

# Coefficients:
#                  Estimate Std. Error z value Pr(>|z|)
#   (Intercept)     -0.3778     0.1175  -3.214 0.001309 **
#   ConditionHigh  -0.6682     0.1765  -3.787 0.000153 ***
#   ConditionLow    0.9100     0.1677   5.427 5.73e-08 ***
#   ---
#   Signif. codes:  0 '***' 0.001 '**' 0.01 '*' 0.05 '.' 0.1 ' ' 1
#
# (Dispersion parameter for binomial family taken to be 1)
#
# Null deviance: 1231.1  on 899  degrees of freedom
# Residual deviance: 1144.6  on 897  degrees of freedom
# AIC: 1150.6
```

CODE BLOCK 36 Modeling Responses by Condition

and 0.63, respectively. These numbers make sense once we look at the percentages in Fig. 7.5.

Much like the models in chapter 6, we could present our results in a table (e.g., Table 6.2) or in a figure (e.g., Fig. 6.6)—we will explore both options later for more complex models. Notice that in the present example we are only entertaining a single predictor of low attachment, namely, Condition. In other words, we are completely ignoring other variables that can potentially be important here—indeed, treating all participants, both native speakers and learners, as a single group makes very little sense given what we know about second language acquisition. But don't worry: this model is here simply to show you how a simple logistic regression is interpreted when we have a categorical predictor (cf. fit_glm1 in §7.2.1). Before we move to our next example, let's see how we could report these results.

REPORTING RESULTS

A logistic model shows that Condition has a significant effect on the probability of choosing low attachment. Relative to our baseline, NoBreak, both high and low breaks affect our participants' preference for low attachment. Having a high break has a negative effect ($\hat{\beta} = -0.67, p < 0.001$), and having a low break has a positive effect ($\hat{\beta} = 0.91, p < 0.0001$).[7]

7.2.3 Do Proficiency *and* Condition *Affect Responses?*

We will now run a logistic model that includes both Condition and Proficiency as predictors of low attachment responses. The factor Proficiency has three levels, which ordered alphabetically are Adv, Int, and Nat. Fortunately, native speakers (our controls) are coded for this variable, which means that by examining Proficiency we will be able to compare not only intermediate to advanced learners of English but also L2ers to native speakers.

In R, our model can be simply represented as Low ~ Condition + Proficiency. Here, we have two categorical predictors, each of which has three levels. We already know what the reference level is for Condition, since we manually set it to NoBreak in code block 36. How about Proficiency? By default, it's Adv, as it's the first level ordered alphabetically. Is that ideal? Probably not. It would make more sense to use our controls (native speakers) as our reference level, so we can compare both intermediate and advanced learners to them. We will do that right before running the model. Before we get there, let's inspect Fig. 7.6, which plots exactly the same variables we will model later.

Fig. 7.6 plots Condition on the x-axis and uses different shades of gray to represent the different levels of Proficiency. Notice that the shades of gray

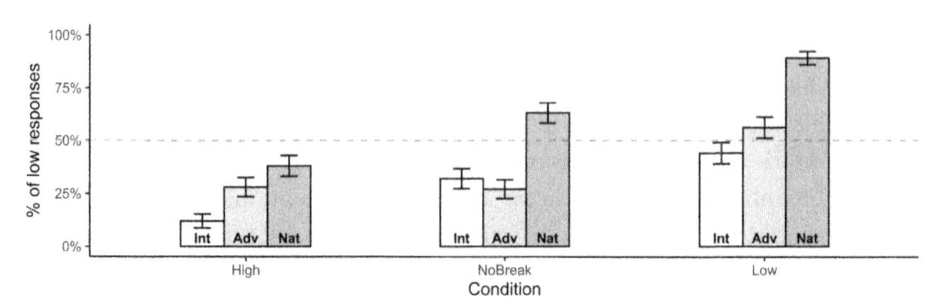

FIGURE 7.6 Preference for Low Attachment by Condition and Proficiency

chosen match the proficiency levels, such that a darker shade represents a more proficient group (native speakers being represented by the darkest shade). In addition, the order of presentation of the proficiency levels also makes sense, that is, intermediate before advanced learners. Finally, even though the fill color of the bars represent proficiency, we don't see a key in the figure. Instead, the actual levels of Proficiency are displayed at the base of each bar. The combination of the labels and intuitive shades of gray makes it easy to quickly understand the patterns in the figure—the code to generate Fig. 7.6 is shown in code block 37, since it has some specific lines that we haven't discussed yet.

We already know the overall pattern involving Condition, but Fig. 7.6 shows the importance of Proficiency on top of Condition. First, let's focus on native speakers, that is, the darkest bars. Recall that English is expected to favor low attachment in general. As a result, it's not a surprise that in the NoBreak (i.e., neutral) condition natives speakers prefer low attachment more than 50% of the time. If you move left, to Condition = High, the bar goes down for native speakers, and if you go right, to Condition = Low, it goes up—both relative to NoBreak, which is our baseline here again.

Learners in the data clearly disprefer low attachment overall—recall that Spanish is expected to favor high attachment in general: we can see that for NoBreak the bars for intermediate and advanced learners are both below 50%. In fact, the only time a bar is above 50% for a non-native group is when Condition = Low and Proficiency = Adv. In other words, advanced English learners seem to prefer low to high attachment given the right condition (a low break in the stimuli).

As we transition from our figure to our model, it's important to remember what exactly is our intercept here, that is, what our reference level is for both predictor variables. As per our discussion earlier, we will set our reference level for Proficiency to native speakers (Nat)—we have already set our reference level for Condition (NoBreak). Therefore, our intercept here represents native speakers in the NoBreak condition.

Before we actually run our model, 7.5 shows how we can represent it mathematically. We have two $\hat{\beta}$ for Condition and two $\hat{\beta}$ for Proficiency. Because

R code

```
# Remember to add this code block to plotsCatModels.R
source("dataPrepCatModels.R")
library(scales) # to add percentages on y-axis using percent_format() below

# Order levels of Proficiency: Int, Adv, Nat
rc = rc %>%
  mutate(Proficiency = factor(as.factor(Proficiency), levels = c("Int", "Adv", "Nat")))

# Make figure:
ggplot(data = rc, aes(x = Condition, y = Low,
                      fill = Proficiency, label = Proficiency)) +
  geom_hline(yintercept = 0.5, linetype = "dashed", color = "gray") +
  stat_summary(geom = "bar",
               alpha = 0.5, width = 0.5,
               color = "black",
               position = position_dodge(width = 0.5)) +
  stat_summary(geom = "errorbar", width = 0.2,
               position = position_dodge(width = 0.5)) +
  theme_classic() +
  geom_text(data = rc %>% filter(Low == 0) %>% mutate(Low = Low + 0.04),
            position = position_dodge(width = 0.5),
            size = 3, fontface = "bold") +
  scale_x_discrete(limits = c("High", "NoBreak", "Low")) +
  scale_fill_manual(values = c("white", "gray60", "gray50")) +
  scale_y_continuous(labels = percent_format()) + # This requires the scales package!
  labs(y = "% of low responses") +
  theme(legend.position = "none")

# ggsave(file = "figures/condition-prof-low-barplot.jpg",
#        width = 8, height = 2.5, dpi = 1000)
```

CODE BLOCK 37 Code for Fig. 7.6: Bar Plot and Error Bars (Three Variables)

both variables are categorical, we can turn them on and off again—the same situation we discussed in §7.2.2. For example, suppose a participant is an intermediate learner, and we're modeling a stimulus in the high break condition. In that case, our model would be defined as $P(Low = 1) = logit^{-1}(\hat{\beta}_0 + \hat{\beta}_{high} \cdot 1 + \xcancel{\hat{\beta}_{low} \cdot 0} + \hat{\beta}_{int} \cdot 1 + \xcancel{\hat{\beta}_{adv} \cdot 0})$. Naturally, if the item we're modeling is in the NoBreak condition and our participant is a native speaker, our model would simply be $P(Low = 1) = logit^{-1}(\hat{\beta}_0)$.

$$P(Low = 1) = logit^{-1}(\hat{\beta}_0 + \overbrace{\hat{\beta}_{high} x_{i_{high}} + \hat{\beta}_{low} x_{i_{low}}}^{\text{Condition}} + \overbrace{\hat{\beta}_{int} x_{i_{int}} + \hat{\beta}_{adv} x_{i_{adv}}}^{\text{Proficiency}}) \tag{7.5}$$

Code block 38 fits our model (lines 1–2 set Nat as the reference level for Proficiency).[8] Let's first look at our intercept ($\hat{\beta}_0 = 0.53, p < 0.001$). We haven't discussed null hypotheses in a while, but you should remember that the null hypothesis for every estimate in our model is that its value is zero. The only difference here is that zero means "zero log-odds", which in turn means 50% probability (see Table 7.1). This makes sense: a 50% probability is no better than chance. The intercept here tells us the log-odds of choosing low attachment for native speakers in the condition NoBreak. The estimate is positive, which means $P(Low = 1) > 0.5$. If you look back at Fig. 7.6, you will see that this makes perfect sense: the bar for native speakers in the NoBreak condition is above 50%.

If we simply look at the signs of our estimates, we can see that relative to our intercept: (i) high breaks lower the log-odds, and (ii) low breaks raise the log-odds of choosing a low attachment response—these effects shouldn't be surprising given our discussion about fit_glm2. Let's focus specifically on Proficiency effects. Relative to native speakers (our intercept), both advanced and intermediate learners choose low attachment *less* frequently—we can see that in Fig. 7.6. Therefore, it's not surprising that the log-odds of both Adv and Int are negative in fit_glm3.

Finally, we could also generate a table with the $P(Low = 1)$ for all possible combinations of variables: all three conditions for each of the three proficiency groups in our data. The table would contain nine rows and could be generated with the predict() function. The predicted probabilities in question would mirror the percentages we observe in Fig. 7.6.

How does the model in question compare with fit_glm2? Intuitively, fit_glm3 is expected to be better, since it includes a predictor that statistically affects our response variable. If you look back at code block 36, you will see that the AIC for fit_glm2 is 1150.6. For fit_glm3, we see in code block 38 that the AIC is 1067.3. The lower AIC value for fit_glm3 tells us that it is the better fit of the two models in question. We could also repeat the steps in code block 35 to determine the accuracy of both models and compare them directly.

R code

```
1  # Remember to add this code block to logModels.R (run code block 36 to relevel Condition)
2  rc = rc %>%
3    mutate(Proficiency = relevel(as.factor(Proficiency), ref = "Nat"))
4
5  fit_glm3 = glm(Low ~ Condition + Proficiency, data = rc, family = "binomial")
6  summary(fit_glm3)
7
8  #    Coefficients:
9  #                     Estimate Std. Error z value Pr(>|z|)
10 #    (Intercept)        0.5302     0.1609   3.295 0.000986 ***
11 #    ConditionHigh     -0.7443     0.1868  -3.984 6.78e-05 ***
12 #    ConditionLow       1.0118     0.1779   5.688 1.28e-08 ***
13 #    ProficiencyAdv    -1.2142     0.1814  -6.695 2.16e-11 ***
14 #    ProficiencyInt    -1.6020     0.1878  -8.532  < 2e-16 ***
15 #    ---
16 #    Signif. codes:  0 '***' 0.001 '**' 0.01 '*' 0.05 '.' 0.1 ' ' 1
17 #
18 # (Dispersion parameter for binomial family taken to be 1)
19 #
20 # Null deviance: 1231.1  on 899  degrees of freedom
21 # Residual deviance: 1057.3  on 895  degrees of freedom
22 # AIC: 1067.3
```

CODE BLOCK 38 Running a Logistic Regression with Two Categorical Predictors

REPORTING RESULTS

A logistic model shows that both Condition and Proficiency are significant predictors of participants' preference for low attachment in their responses. More specifically, both high ($\hat{\beta} = -0.74, p < 0.001$) and low ($\hat{\beta} = 1.01, p < 0.0001$) breaks in the stimuli significantly affect the probability that a participant will choose low attachment in the data (relative to the NoBreak condition, our intercept in the model). As for proficiency, both advanced ($\hat{\beta} = -1.21, p < 0.0001$) and intermediate ($\hat{\beta} = -1.60$, $p < 0.0001$) learners choose low attachment significantly less frequently than native speakers, consistent with previous studies and with what we observe in Fig. 7.6.

7.2.4 Do Proficiency *and* Condition *Interact?*

So far we have assumed that Condition and Proficiency do not interact as variables in our model. In other words, we assume that the effects of Condition are the same for all three groups of participants we have been examining. However, we have reason to believe an interaction exists here: if you look back at Fig. 7.6, you will notice that the patterns are not all identical across all three sets of bars. For example, for the NoBreak condition, intermediate and advanced learners pattern together (overlapping error bars), but for the other two conditions they don't. This could mean a significant interaction exists between the two variables in question. That's what we'll test with our final model in this chapter, fit_glm4, in code block 39. Because we have a two-way interaction this time, it will be more complicated to examine the interpretation of the model in detail (you may recall the discussion from §6.2.4). Naturally, you don't need to report a comprehensive interpretation of the results in your paper, but we will do that here to understand exactly what an interaction means again.

This model, fit_glm4, will be the most complex logistic model we have run so far, in that it has an intercept, four simple terms, and four interacting terms in its specification. In R, our model is specified as Low ~ Proficiency * Condition, which is fairly easy to remember—it's also very concise when compared to its actual mathematical specification, shown in 7.6 (H = high, L = low, I = intermediate, A = advanced).

The first line of our model in 7.6 is exactly the same as before (see 7.5). The second line is simply the continuation of the first and includes all four interaction terms, which are basically the combinations of Condition and Proficiency: high-intermediate, low-intermediate, high-advanced, and low-advanced.

Fortunately, we will never have to use 7.6 manually: R will do all that for us.

$$P(Low = 1) = logit^{-1}(\hat{\beta}_0 + \overbrace{\hat{\beta}_H x_{i_H} + \hat{\beta}_L x_{i_L}}^{\text{Condition}} + \overbrace{\hat{\beta}_I x_{i_I} + \hat{\beta}_A x_{i_A}}^{\text{Proficiency}} + \underbrace{\hat{\beta}_{H*I} x_{i_H} x_{i_I} + \hat{\beta}_{L*I} x_{i_L} x_{i_I} + \hat{\beta}_{H*A} x_{i_H} x_{i_A} + \hat{\beta}_{L*A} x_{i_L} x_{i_A})}_{\text{Interaction}} \quad (7.6)$$

Code block 39 shows the output of fit_glm4. You will notice that the reference level of Condition is NoBreak, and the reference level of Proficiency is Nat. If your output looks different, you should rerun code blocks 36 and 38, which relevel both variables—all three code blocks should be in logModels.R.

As usual, let's start general and get more specific as we discuss our effects here. First, notice that all main effects are significant, including the intercept. Second, two interactions are also significant (lines 12 and 13 in our code block). So we were right to suspect that an interaction between Condition and Proficiency existed in our data given Fig. 7.6—one more reason that visualizing our patterns is essential before we start exploring our models. Next, notice that the AIC of this model is 1060.3, the lowest number yet. This already tells us that fit_glm4 is the best model so far (excluding fit_glm1, which modeled a different variable, i.e., L2er).

Intercept. The intercept here is again the predicted log-odds of Low = 1 when all other variables are set to zero. What does that mean? Well, here, that's when Condition = NoBreak and Proficiency = Nat, since that's what the intercept represents in our model. Notice that the estimate is positive ($\hat{\beta}_0 = 0.53, p < 0.05$), which means the probability of choosing low attachment for native speakers in the NoBreak condition is above chance (50%)—again, this should not be surprising given our discussion so far (and given Fig. 7.6).

ConditionHigh. This is the predicted log-odds of choosing low attachment in condition High *assuming* that Proficiency = Nat. That the estimate is negative ($\hat{\beta} = -1.02$) is not surprising, since we know that native speakers disprefer low attachment in the high condition (relative to the NoBreak condition).

ConditionLow. This is the predicted log-odds of choosing low attachment in condition Low *assuming* that Proficiency = Nat. That the estimate is positive is not surprising either, since we know that native speakers prefer low attachment in the low condition (relative to the NoBreak condition). As you can see, so far we've been interpreting our estimates assuming the native speaker group—recall our discussion back in chapter 6.

ProficiencyInt. This is the predicted log-odds of choosing low attachment for intermediate learners *assuming* the NoBreak condition. Notice a pattern here? When we interpret one variable, we assume the other one is held at

R code

```
fit_glm4 = glm(Low ~ Condition * Proficiency, data = rc, family = "binomial")
summary(fit_glm4)

#   Coefficients:
#                                Estimate Std. Error z value Pr(>|z|)
#   (Intercept)                    0.5322     0.2071   2.570  0.01018 *
#   ConditionHigh                 -1.0218     0.2921  -3.498  0.00047 ***
#   ConditionLow                   1.5585     0.3808   4.092 4.27e-05 ***
#   ProficiencyInt                -1.2860     0.2981  -4.314 1.60e-05 ***
#   ProficiencyAdv                -1.5268     0.3060  -4.990 6.05e-07 ***
#   ConditionHigh:ProficiencyInt  -0.2169     0.4754  -0.456  0.64822
#   ConditionLow:ProficiencyInt   -1.0459     0.4812  -2.173  0.02975 *
#   ConditionHigh:ProficiencyAdv   1.0719     0.4309   2.488  0.01286 *
#   ConditionLow:ProficiencyAdv   -0.3227     0.4862  -0.664  0.50679
# ---
#   Signif. codes:  0 '***' 0.001 '**' 0.01 '*' 0.05 '.' 0.1 ' ' 1
#
# (Dispersion parameter for binomial family taken to be 1)
#
# Null deviance: 1231.1  on 899  degrees of freedom
# Residual deviance: 1042.3  on 891  degrees of freedom
# AIC: 1060.3
```

CODE BLOCK 39 Running a Logistic Regression with an Interaction

zero, which means we assume the level represented by the intercept. The estimate is negative ($\hat{\beta} = -1.29$), which simply captures the observation that intermediate learners disprefer low attachment relative to native speakers in the NoBreak condition—make sure you return to Fig. 7.6 to see how our interpretation matches the patterns in the figure.

ProficiencyAdv. This is the predicted log-odds of choosing low attachment for advanced learners *assuming* the NoBreak condition. The estimate is also negative, which captures the observation that advanced learners disprefer low attachment relative to native speakers in the NoBreak condition.

ConditionLow:ProficiencyInt. Examine only the bars for Condition = Low in Fig. 7.6. Now examine the bar for intermediate learners, and compare it to that of native speakers. If you compare the difference between intermediate learners and native speakers in the low condition and in the NoBreak condition, you will notice that the difference is more pronounced in the low condition. Our model is basically telling us that this difference in magnitude is significant ($\hat{\beta} = -1.05, p < 0.05$). Again: this is telling us that the relationship between Condition and Proficiency is *not* constant in the data. More specifically, we see that the difference in response patterns between intermediate learners and native speakers is not the same when we compare the NoBreak and Low conditions. Indeed, the difference intensifies when Condition = Low, since the main effect for ProficiencyInt is already negative (see earlier).

CONDITIONHIGH:PROFICIENCYADV. Let's compare advanced learners to native speakers, when we go from NoBreak to High—again, check Fig. 7.6. In the high condition, advanced learners are not extremely different from native speakers: they are both under 50%, and their error bars in the figure almost touch, so these two groups are not too far from each other. Now look at the NoBreak condition: native speakers are way above 50%, while advanced learners are at 25%. In other words, as we go from NoBreak to High, the difference between advanced learners and native speakers is *reduced*—hence the positive sign in our estimate. Notice that this is the opposite situation to what we discussed in the interaction earlier: there, the difference intensified as we went from NoBreak to Low; here, it weakened, hence the positive effect ($\hat{\beta} = 1.08, p < 0.05$)—recall again that the main effect of ProficiencyAdv is negative.

The significant interactions in our model are not easy to interpret: it takes some practice to get used to interactions in statistical models. But an easier way to understand the intuition behind what's happening is to step back and simply observe Fig. 7.6. Our significant interactions simply tell us that the effect of condition is not the same for all three groups of proficiency we've been discussing. Instead, how much it affects participants' responses depends in part on which proficiency group they belong to. What this all captures is that the different heights of our bars do not follow the exact same pattern across all three conditions on the *x*-axis in Fig. 7.6—and that's what we observe.

Next, let's report our results. How much detail you want to provide depends on a series of factors. Here I provide a general picture and then discuss the case of native speakers as an example. You could also discuss the learners, as well as the interaction we examined earlier. Ultimately, given that we have a more complex model now, you should definitely refer to a table or figure: going over all the significant estimates in detail is likely not as effective as you think: it's simply too much information. A table is provided later, and a figure will be discussed next.

REPORTING RESULTS

A statistical model confirms that participants' preference for low attachment is significantly affect by Condition, Proficiency, and the interaction of both variables—estimates are provided in Table 7.2. Consistent with the literature, native speakers of English favor low attachment in the NoBreak ($\hat{\beta}_0 = 0.532, p = 0.01$) and Low ($\hat{\beta} = 1.559, p < 0.001$) conditions. For the High condition, on the other hand, native speakers disfavor low attachment ($\hat{\beta} = -1.022, p < 0.001$). This shows that a prosodic effect can impact speakers' interpretation of relative clause heads in ambiguous sentences. Finally, learners in this (hypothetical) study show the effects of their L1 in their response patterns: they overall prefer low attachment *less* than the native speakers in the study (see estimates in Table 7.2).

In chapter 6, we also saw that we can use figures to present/report a model's estimates. To produce a figure with the estimates for fit_glm4, we take essentially the same steps described in code blocks 30 and 31, back in chapter 6. Ultimately, we are doing exactly the same thing: we extract the estimates from our fit, then we extract the 95% confidence intervals and use that to create a figure. The results of fit_glm4 are shown in Fig. 7.7—you can compare the figure to Table 7.2 to see which one you find more intuitive (they show exactly the same thing, of course).

Let's briefly discuss what's different about Fig. 7.7. First, the names of the predictors have been shortened, so our *y*-axis doesn't need too much space. Second, here we're using point range as opposed to error bars (cf. Fig. 6.6). Which one you prefer is a matter of taste: point ranges will indicate where the estimate is with a point in the figure. Third, two of our point ranges have dashed lines. Those two confidence intervals cross zero, which means they are not significant ($p > 0.05$). By using solid lines for significant effects and dashed lines for non-significant effects, we make it easier for the reader to immediately see which effects are statistically confirmed by our model.

How do we use two line types for significance? Code block 40 shows you how to do it manually. Notice that lines 20–22 create a new variable called Sig, which will be set to Yes in two situations: (i) when the lower and upper bounds of the 95% confidence interval are above zero, or (represented by "|") (ii) when both are below zero. Otherwise, the interval will include zero, in which case we set Sig to No. Later, in the figure (line 25), we

TABLE 7.2 Model Estimates and Associated Standard Errors, *z*-values, and *p*-values

	Estimate	*Std. Error*	*z-value*	*Pr(>|z|)*
(Intercept)	0.532	0.207	2.570	0.01
ConditionHigh	−1.022	0.292	−3.498	< 0.001
ConditionLow	1.559	0.381	4.092	< 0.001
ProficiencyInt	−1.286	0.298	−4.314	< 0.001
ProficiencyAdv	−1.527	0.306	−4.990	< 0.001
ConditionHigh:ProficiencyInt	−0.217	0.475	−0.456	0.648
ConditionLow:ProficiencyInt	−1.046	0.481	−2.173	0.030
ConditionHigh:ProficiencyAdv	1.072	0.431	2.488	0.013
ConditionLow:ProficiencyAdv	−0.323	0.486	−0.664	0.507
				AIC = 1060.3

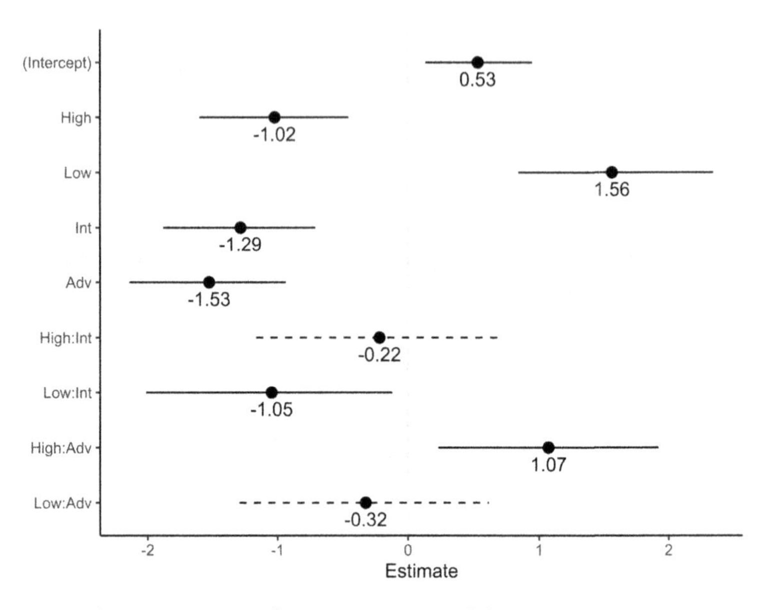

FIGURE 7.7 Plotting Estimates from a Logistic Model

can use different line types in our geom_pointrange() to represent the two levels of Sig. Finally, in line 34, we manually choose which line types we want to represent both No and Yes—in that order, since R orders levels alphabetically.

Let's now briefly go over the goodness of fit of fit_glm4. We have already established that the model's AIC is the lowest of the three models so far, which tells us that it has a better fit than the other two models. In addition, we can use the anova() function to compare the two models using a likelihood ratio test: anova(fit_glm3, fit_glm4, test = "LRT"), which will reveal that fit_glm4 has a significantly better fit (i.e., it has lower deviance from the data relative to fit_glm3). As a result, this would be the model to report. None of this is surprising: the *p*-values we get from summary(fit_glm4) are computed using Wald tests, which basically compare a model with predictor X to a model without said predictor to check whether X is actually statistically relevant. We've seen that the interaction between Condition and Proficiency is significant, so it makes sense that a model without said interaction would be less optimal than a model *with* the interaction.

But how accurate is fit_glm4? In other words, what percentage of the data modeled does the model predict accurately? You may recall that we examined the accuracy of fit_glm1, which predicted the probability of being a learner based on a participant's reaction time—see code block 35. The answer here is 70%: fit_glm4 can accurately predict 70% of the responses in the data—

R code

```
# Plot estimates and confidence intervals for fit_glm4
# Prepare data:
glm4_effects = tibble(Predictor = c("(Intercept)", "High",
                                    "Low", "Adv", "Int",
                                    "High:Adv", "Low:Adv",
                                    "High:Int", "Low:Int"),
                      Estimate = c(coef(fit_glm4)[[1]],   # Intercept
                                   coef(fit_glm4)[[2]],   # High
                                   coef(fit_glm4)[[3]],   # Low
                                   coef(fit_glm4)[[4]],   # Adv
                                   coef(fit_glm4)[[5]],   # Int
                                   coef(fit_glm4)[[6]],   # High:Adv
                                   coef(fit_glm4)[[7]],   # Low:Adv
                                   coef(fit_glm4)[[8]],   # High:Int
                                   coef(fit_glm4)[[9]]),  # Low:Int
                      l_CI = c(confint(fit_glm4)[1:9]),   # lower CI
                      u_CI = c(confint(fit_glm4)[10:18])) # upper CI

# Add binary column for sig vs. not sig:
glm4_effects = glm4_effects %>%
  mutate(Sig = ifelse(l_CI > 0 & u_CI > 0 | l_CI < 0 & u_CI < 0,
                      "yes", "no"))

ggplot(data = glm4_effects, aes(x = Predictor, y = Estimate)) +
  geom_pointrange(aes(ymin = l_CI, ymax = u_CI, linetype = Sig)) +
  coord_flip() + theme_classic() +
  scale_x_discrete(limits = rev(c("(Intercept)", "High", "Low",
                                  "Int", "Adv", "High:Int", "Low:Int",
                                  "High:Adv", "Low:Adv"))) +
  geom_text(aes(label = round(Estimate, digits = 2)),
            position = position_nudge(x = -0.3, y = 0)) +
  geom_hline(yintercept = 0, linetype = "dashed", alpha = 0.1) +
  labs(x = NULL) + theme(legend.position = "none") +
  scale_linetype_manual(values = c("dashed", "solid"))

# Save plot:
# ggsave(file = "figures/model-estimates.jpg", width = 6, height = 4.5, dpi = 1000)

# Alternatively: install sjPlot package
# library(sjPlot)
# plot_model(fit_glm4, show.intercept = TRUE) + theme_classic()
```

CODE BLOCK 40 Creating a Plot for Model Estimates

this doesn't necessarily mean that if we ran the model on new data it would have the same accuracy (see discussion on the accuracy of fit_glm1 in §7.2.1). The code to calculate the accuracy of fit_glm4 is shown in code block 41.

Finally, we should also spend some time on model diagnostics once we have run and selected our model of choice. In chapter 6, we saw that plotting the residuals of a model is essential for model diagnostics. For logistic regressions, however, only plotting residuals will not be very helpful. Instead, you should use the binnedplot() function in the arm package. For example, for fit_glm4, we could use the function in question to plot expected values against average residuals. This is known as a *binned residual plot* and can be generated by running binnedplot(predict(fit_glm4), residuals(fit_glm4)). Most points in the plot should fall within the confidence limits (gray lines) shown—see Gelman and Hill (2006, pp. 97–98) for more details on binned plots.

7.3 Summary

In this chapter we discussed the basic characteristics and assumptions of logistic models, and we examined different examples of such models using our hypothetical study on relative clauses. We have also discussed how to report and present our results, how to compare different models, and how to calculate the accuracy of a model using the data modeled as the model's input. It's time to review the most important points about logistic models.

- Logistic regressions are used to model a binary response variable. As a result, we predict the probability of an event happening (*vs.* not happening).
- Probabilities can only go from 0 to 1, so we have to adjust our linear model from chapter 6 accordingly: instead of modeling probabilities, which are not linear, or odds, which have asymmetrical magnitudes, we model the log-odds of a response. Therefore, the estimates in a logistic regression are given in log-odds.
- We can easily transform log-odds to odds (exp()) or probabilities (invlogit () in the arm package).

R code
```
1  # Measuring accuracy of fit_glm4:
2  fit_glm4_accuracy = rc %>%
3    dplyr::select(Condition, Proficiency, Low) %>%
4    mutate(pLow = predict(fit_glm4,
5                          newdata = tibble(Condition = Condition,
6                                           Proficiency = Proficiency),
7                          type = "response"),
8           LowBin = ifelse(pLow > 0.5, 1,
9                           ifelse(pLow < 0.5, 0, NA)),
10          Accuracy = ifelse(LowBin == Low, 1, 0))
11
12 fit_glm4_accuracy %>%
13   summarize(Correct = sum(Accuracy)/nrow(.))
14
15 # A tibble: 1 x 1
16 #     Correct
17 #       <dbl>
18 #   1   0.697
```

CODE BLOCK 41 Assessing Model's Accuracy: fit_glm4

- We can choose our models based on the lowest AIC value they achieve: lower AIC values indicate better fits.
- It's also possible to compute the accuracy of a model, which tells us how much of the data modeled is accurately predicted by the model being fit to the data.
- Finally, we saw how to report our results in text and to present our estimates in a table or in a figure.
- Most of the coding used for linear regressions are applicable to logistic regressions, since both types of models belong to the same family (generalized linear models). Indeed, even the interpretation of our model estimates is very similar to what we have in linear regressions—once you get used to log-odds in logistic models, of course. In other words, if you know how to interpret an interaction in a linear regression, you also know how to do that in a logistic regression *as long as* you understand log-odds. Likewise, if you know how to rescale variables for a linear regression, you also know how to do that for a logistic regression.
- Finally, you should use binned residual plots for model diagnostics. Such plots are a quick and intuitive way to check whether the model you plan to report is actually working as it should.

7.4 Exercises

Ex. 7.1 Basics

1. The estimates here are given in log-odds, as usual. Transform them into odds and probabilities using R.
    - 0.74, 0.1, 1.8

2. If an estimate is 0.57 and its standard error is 0.18, what is its 95% confidence interval? Is the estimate statistically significant?

Ex. 7.2 To Rescale or Not to Rescale?

1. Would it be a good idea to rescale our predictors in fit_glm3? Why (or why not)?
2. How would rescaling our predictors affect our interpretation of the intercept?

Ex. 7.3 More Predictors

1. Does reaction time correlate with participants' attachment preferences? Create a figure you consider appropriate to examine the question. You may or may not add Proficiency to the figure. *Hint:* You can create a box

plot or a bar plot with error bars with Response on the x-axis and then use coord_flip() to make your response variable appear on the y-axis.

2. Run a logistic model that accompanies the figure you just created—that is, where you add reaction time as a predictor. Would we want to rescale RT? Why? Why not? Do the results surprise you given the figure you just created?

Ex. 7.4 Accuracy, Diagnostics, and Predictability

1. Calculate the accuracy of fit_glm3. How different is it from the accuracy of fit_glm4? Are you surprised by the result? Why?
2. Create a binned residual plot for fit_glm4. Ideally, we should see 95% of the points plotted within the error bounds (confidence limits). Does that happen here?
3. Using fit_glm4, what's the predicted probability of choosing low attachment for the following participants:

    a) An Int learner in the High condition
    b) An Adv learner in the Low condition
    c) A Nat speaker in the NoBreak condition

Notes

1. What 1 and 0 represent is entirely up to you. You could assign 0 to Spanish speakers, in which case our figure would plot the probability that participant i is a native speaker of English. We will return to this discussion later, in §7.2.
2. More specifically, we'll be using the Bernoulli distribution, which in our data represents the probability of being a learner for each individual data point (i.e., each trial) in the data. A binomial distribution is the sum of multiple trials that follow a Bernoulli distribution.
3. If you don't remember the hypothetical study in question, return to chapter 4, where the data is first presented.
4. Not a t-value, which we saw in our linear models in chapter 6.
5. You should be able to produce the figure in question by going back to chapter 4—remember to add the code to plotsCatModels.R and to load the scales package to add percentages to the y-axis. The code for the figure can be found in the files that accompany this book.
6. Notice that the labels on the x-axis have been reordered and that the x-axis here plots variable Low, created in code block 33. When we have a column with 0s and 1s, stat_summary() will automatically give us proportions.
7. We could also report confidence intervals by using the confint() function, or the change in odds by taking the exponential of the log-odds, as we did in our discussion earlier.
8. We have already changed the reference level of Condition back in code block 36, which at this point you have already run.

8

ORDINAL REGRESSION

In chapter 6, we examined linear regressions, which are used to model a continuous response variable. In chapter 7, we examined logistic regressions, which help us model a binary response variable. As denoted by the title of the present chapter, we now turn to *ordinal* variables, our third and last data type. For example, you may want your participants to choose how natural a particular sentence sounds or how certain they are about their responses or how satisfied they are with their learning after a given period of time. All these questions are typically examined with scales—quite often Likert scales, named after American psychologist Rensis Likert.

A Likert scale is balanced and symmetrical. For example, suppose we present our participants with the statement "Learning a second language is more challenging to adult learners". We then ask them to use a 5-point Likert scale with the following categories: STRONGLY AGREE, AGREE, NEUTRAL, DISAGREE, STRONGLY DISAGREE. Here, we have complete symmetry (cf. discussion in §4.2): the categories at the end-points of the scale have opposite meanings. Likewise, AGREE is the opposite of DISAGREE, and the neutral point is characterized by an unambiguously neutral label, that is, NEUTRAL. You may also see studies that only label end-points—an example is shown in Fig. 8.1.

To analyze ordinal data, in this chapter we will see how to run and interpret ordinal models, also known as cumulative link models (Agresti 2002). More specifically, we will focus on logit ordered models, so a lot of what we will discuss can be transferred from logistic regressions from chapter 7.[1]

As usual, before we get started, let's create a new script—you should place these scripts in the Frequentist folder we created in chapter 6, so all the files associated with our models will be in the same directory. You should already

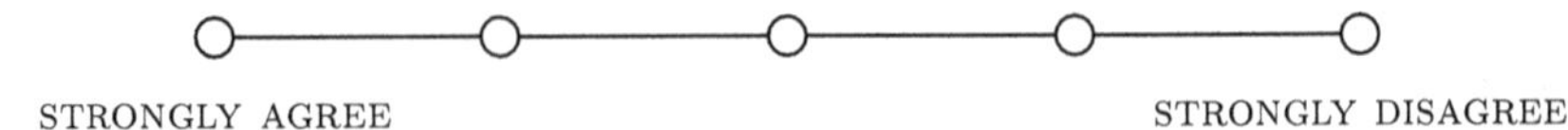

FIGURE 8.1 A Typical Scale with Labeled End-Points

have a copy of rClauseData.csv in there too—we'll be running our models on the dataset used in chapters 4 and 5. The script we will create is ordModels.R. Because we already have Frequentist.RProj in the same directory, we can use a single R Project to manage all the files connected to the Frequentist statistical models in this book. We will still use dataPrepCatModels.R and plotsCatModels.R, created in chapter 7.

> **DATA FILE**
>
> In this chapter (§8.2), we will use rClauseData.csv again. This file simulates a hypothetical study on relative clauses in second language English.

8.1 Introduction

The type of ordinal model we will examine in this chapter is essentially a logistic regression adapted to deal with three or more categories in a response variable. More specifically, we will focus on *proportional odds models*—see Fullerton and Xu (2016, ch. 1) for a typology of ordered regression models. While in a logistic regression we want to model the probability that a participant will respond "yes" (as opposed to "no"), in an ordinal logistic regression we want to model a *set of probabilities*. For example, let's assume that we have three categories in a scale where participants had to choose their level of certainty: NOT CERTAIN, NEUTRAL, and CERTAIN. What we want to model is the probability that a given participant is *at least* NOT CERTAIN—which in this case simply means "not certain", since there's no other category below NOT CERTAIN. We also want to model the probability that a participant is *at least* NEUTRAL, which in this case includes the probability of NOT CERTAIN *and* NEUTRAL—so we are essentially asking for the *cumulative* probability of a given category in our scale. Because our hypothetical scale here only has three categories, we only need two probabilities: to find out the probability that a participant is CERTAIN we simply subtract the cumulative probability of being (at least) NEUTRAL from 1.

The structure of an ordinal model is shown in 8.1, where we only have a single predictor variable. Here, for a scale with J categories, we want to model the cumulative probability that a response is less or equal to category $j = 1, \ldots, J - 1$. Notice that our intercept, now represented as $\hat{\tau}_j$ (the Greek letter tau), also has a specification for category j. That's because in an ordinal regression we have multiple intercepts ($J - 1$ to be exact), which represent cut-points or thresholds in our scale. Therefore, if our scale has three categories, our model will give us two thresholds. We normally don't care about τ, but we will see how it affects our interpretation soon. Finally, notice the minus sign before $\hat{\beta}_1$ in 8.1. The reason for subtracting our coefficients from $\hat{\tau}_j$ is simple: an association with higher categories on our scale entails *smaller* cumulative probabilities for lower categories on our scale.

The intuition behind an ordinal model is relatively straightforward, especially if you're mostly interested in a broad interpretation of the results. If you wish to have a more detailed interpretation of coefficients and thresholds, then it can take a little longer to get used to these models. The good news is that a broad interpretation is all you need 99% of the time.

$$P(Y \leq j) = logit^{-1}(\hat{\tau}_j - \hat{\beta}_1 x_{i_1}) \tag{8.1}$$

Because we have already covered logistic regressions in chapter 7, let's now turn to some real examples—running some models and interpreting their results will help you understand the subtle differences between logistic and ordinal models. As we will see, ordinal logistic regressions are very similar to logistic regressions. Indeed, if we only had two categories in our scale, an ordinal logistic regression would be identical to a logistic regression. Finally, to run our models, you will need to install the `ordinal` package (Christensen 2019), so go ahead and run `install.packages("ordinal")`.

8.2 Examples and Interpretation

8.2.1 Does Condition *Affect Participants' Certainty?*

Let's begin by opening our `ordModels.R` script, which is where we will run our models—recall that here we will analyze the `rClauseData.csv`, our hypothetical relative clause study. Inside `ordModels.R`, let's call our other script, `dataPrepCatModels.R`, which contains the code in code block 33 back in chapter 7—notice that by sourcing `dataPrepCatModels.R` you are already loading `tidyverse`. By loading `dataPrepCatModels.R`, you are also loading the data and assigning it to a new variable, `rc`—this should all be familiar at this point. If it isn't, you should definitely review chapter 7 before proceeding.

Visualize the top rows of `rc` so you can remember what the dataset looks like. You will notice we have a column called `Certainty`, which until now we haven't analyzed—we did, however, use that column to plot ordinal data back in chapters 4 and 5. The `Certainty` column will be our focus in this chapter, for obvious reasons. As usual, before actually running any models, we should take a look at our data.

The variable Certainty comes from a 6-point certainty scale used by participants to indicate how certain they were about their responses. We will simplify our scale, though: instead of 6 categories, let's derive a 3-point scale from it. Our new scale will mirror the scale discussed earlier, so it will be Not certain if Certainty < 3, Certain if Certainty > 4, and Neutral otherwise. We are essentially compressing our scale into three blocks: [1 2] [3 4] [5 6]. The good news is that this is already done in code block 33, which should be located in dataPrepCatModels.R. Our new variable/column is called Certainty3, so it's easy for us to remember that this is the version of Certainty that has 3, not 6, categories. Finally, notice that lines 28–31 in code block 33 adjust the order of the levels of Certainty3 such that Not Certain < Neutral < Certain—by default, R would order them alphabetically, but we want to make sure R understands that Certain is "greater than" Neutral.

Let's take a look at how the certainty levels of the two groups involved in the study (English and Spanish) vary depending on the condition of our experiment (Low, High, and NoBreak). For now, we will focus only on our target group, namely, the learners (Spanish speakers). So focus on the right facet of Fig. 8.2. We can see that in the high condition (at the bottom of the y-axis), Spanish speakers are certain of their responses more than 50% of the time. In contrast, for the low condition these learners are certain of their responses less than 15% of the time. This is in line with the typological differences between the two languages: while in English low attachment is typically preferred, in Spanish it's high attachment that is favored. Therefore, we can see some influence of the participants' L1 in their certainty levels (recall that the experiment was in English, not Spanish).

Fig. 8.2 seems to have no x-axis label. The reason is simple: the actual percentages have been moved to the bars, so it's very easy to see the values. In addition, the shades of gray make it intuitive to see which bars represent

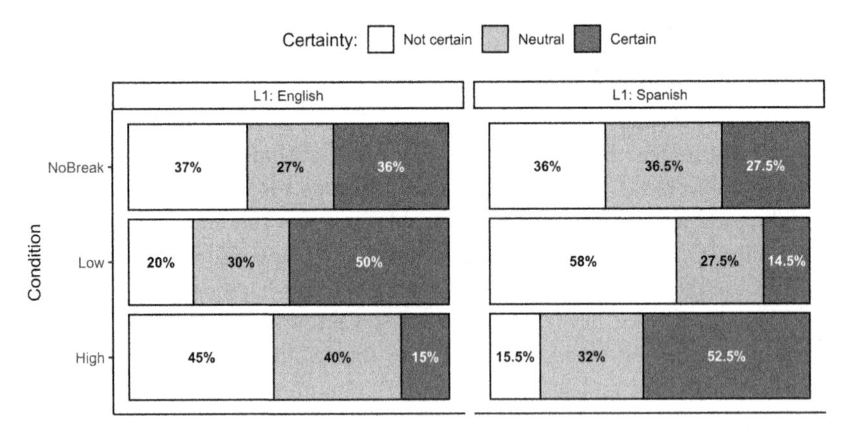

FIGURE 8.2 Certainty Levels (%) by Condition for English and Spanish Groups

Certain and which represent Not certain. The code to produce Fig. 8.2 is shown in code block 42—make sure you place this code inside plotsCatModels.R.

Fig. 8.2 clearly shows a potential effect of Condition on Certainty3, so that will be the focus of our first model: we want to know how the certainty level of our Spanish-speaking participants is affected by the experimental conditions in question. Our model is therefore defined as Certainty3 ~ Condition in R.

Code block 43 runs our ordinal model—you can add this code block to your ordModels.R script. First, we load the script that prepares our data, dataPrepCatModels.R. Next, we load the ordinal package (make sure you have installed it). Lines 7–8 relevel Condition so that NoBreak is our reference level. Line 10 runs the actual model by using the clm() function (*cumulative link model*).[2] Notice that line 10 runs the model on a subset of the data (only Spanish speakers).[3] We're doing that here because we want to start with a simple model, which ignores the differences between the two groups and only focuses on the differences between the conditions for a single group. Thus, fit_clm1 represents only the right facet in Fig. 8.2.

The output of our model, fit_clm1, is shown in lines 14–24. We have our coefficients in lines 16–17 and our thresholds $\hat{\tau}$ in lines 23–24. Recall from the

R code

```
1  # Remember to add this code block to plotsCatModels.R
2  # Calculate proportions:
3  propsOrd = rc %>%
4    group_by(L1, Condition, Certainty3) %>%
5    count() %>%
6    group_by(L1, Condition) %>%
7    mutate(Prop = n / sum(n),
8           Dark = ifelse(Certainty3 == "Certain", "yes", "no"))
9
10 # Make figure:
11 ggplot(data = propsOrd, aes(x = Condition, y = Prop, fill = Certainty3)) +
12   geom_bar(stat = "identity", color = "black") +
13   geom_text(aes(label = str_c(Prop*100, "%"), color = Dark),
14             fontface = "bold", size = 3,
15             position = position_stack(vjust = 0.5)) +
16   facet_grid(~L1, labeller = "label_both") +
17   scale_fill_manual(values = c("white", "gray80", "gray50")) +
18   scale_color_manual(values = c("black", "white"), guide = FALSE) +
19   scale_y_reverse() +
20   coord_flip(ylim = c(1,0)) +
21   theme_classic() +
22   theme(axis.text.x = element_blank(),
23         axis.ticks.x = element_blank(),
24         legend.position = "top") +
25   labs(y = NULL, fill = "Certainty:")
26
27 # ggsave(file = "figures/certainty3.jpg", width = 7, height = 3.5, dpi = 1000)
```

CODE BLOCK 42 Plotting Certainty As a Function of L1

discussion earlier that in an ordinal model we have multiple thresholds (at least two): fit_clm1 has two, given that Certainty3 has three categories ($J = 3$).

The broad interpretation of our estimates ($\hat{\beta}$) is straightforward: condition High (line 16) increases the probability of being certain relative to condition NoBreak ($\hat{\beta} = 1.08$). Conversely, condition Low (line 17) *decreases* the probability of being certain relative to condition NoBreak ($\hat{\beta} = -0.88$). Both conditions have a significant effect on participants' certainty levels, which in turn means that both the effect of High and the effect of Low are significantly different from the effect of NoBreak. Remember: all the estimates in a logit ordered model are given in log-odds (see 8.1), just like in a logistic regression. We could therefore be more specific and say that when Condition = High, the odds of being certain about one's response go up by a factor of $e^{|\hat{\beta}|} = e^{|1.08|} = 2.94$ ("being certain" here simply means "leaning towards the right end point of the scale"). Let's see how we could report these results first, and then we'll explore the interpretation of the model in greater detail.

Most of the functions we discussed back in chapters 6 and 7 are applicable here too. For instance, you can get the confidence intervals of our estimates by running confint(fit_clm1), and you can also manually calculate them using the standard errors in our output (§1.3.4). Likewise, you don't actually need p-values here, since we have both the estimates and their respective standard errors.

R code

```
1  rm(list=ls())
2  source("dataPrepCatModels.R")
3
4  library(arm) # to use invlogit()
5  library(ordinal)
6
7  rc = rc %>%
8    mutate(Condition = relevel(factor(Condition), ref = "NoBreak"))
9
10 fit_clm1 = clm(Certainty3 ~ Condition, data = rc %>% filter(L1 == "Spanish"))
11
12 summary(fit_clm1)
13
14 #   Coefficients:
15 #                  Estimate Std. Error z value Pr(>|z|)
16 #   ConditionHigh    1.0776     0.1904   5.661 1.51e-08 ***
17 #   ConditionLow    -0.8773     0.1920  -4.569 4.90e-06 ***
18 #   ---
19 #   Signif. codes:  0 '***' 0.001 '**' 0.01 '*' 0.05 '.' 0.1 ' ' 1
20 #
21 # Threshold coefficients:
22 #                            Estimate Std. Error z value
23 # Not certain|Neutral         -0.5706     0.1372  -4.159
24 # Neutral|Certain              0.9632     0.1420   6.784
```

CODE BLOCK 43 Modeling Certainty As a Function of Condition

Notice that we didn't care about $\hat{\tau}$ in our broad interpretation earlier, nor did we report what the threshold values are. Let's now spend some time going over a more detailed interpretation of our model, which will require our $\hat{\tau}$ values.

REPORTING RESULTS

An ordinal model confirms that the different conditions in our experiment significantly affect learners' certainty levels. The high attachment condition significantly increases the probability of a learner being more certain about his or her response ($\hat{\beta} = 1.08, p < 0.001$) relative to the condition with no break in it. Conversely, the low attachment condition significantly decreases the probability of a learner being more certain about his or her response ($\hat{\beta} = -0.88, p < 0.001$) relative to the condition with no break in it.[4]

What intercepts mean is essentially constant for all models we discuss in this book. In linear and logistic regressions, the intercept indicated the predicted $\hat{y}$ value when all predictors are zero—the difference, of course, was that in a logistic regression the predicted value is given in log-odds. As mentioned earlier, the ordinal models we are dealing with now are essentially logistic regressions.

If you look back at the output of fit_clm1 in code block 42, you will notice that the values of our thresholds are $\hat{\tau}_1 = -0.57$ and $\hat{\tau}_2 = 0.96$. These values are shown in Fig. 8.3, which overlays our scale categories on top of a normal (Gaussian) distribution. Recall that because we have three categories, we only get two $\hat{\tau}$ values in our model. As you go back to the output of our model, notice that $\hat{\tau}_1$ is the threshold between Not certain and Neutral. Therefore, $\hat{\tau}_1$ represents the log-odds of being at least in category Not certain (or less, which doesn't apply here since this is the first category in the scale).

But which condition are we talking about when we interpret the meaning of $\hat{\tau}_1$ here? The answer is NoBreak, our reference level. Think about it: an intercept gives us the predicted response value assuming that all other terms are set to zero. For categorical predictors, we choose one level of our factor to be our reference. For condition, that level is NoBreak (lines 7–8 in code block 42). So $\hat{\tau}_1$ here essentially gives us the log-odds that a Spanish-speaking participant will choose at least Not certain for the NoBreak condition. If you take the inverse logit of $\hat{\tau}_1$, you will get $P = 0.36$ (invlogit(-0.57) in R using the arm package). That probability is the shaded area under the curve in Fig. 8.3a.

How about $\hat{\tau}_2$? Our second intercept/threshold here simply tells us the cumulative probability of choosing Neutral (or less). The probability is

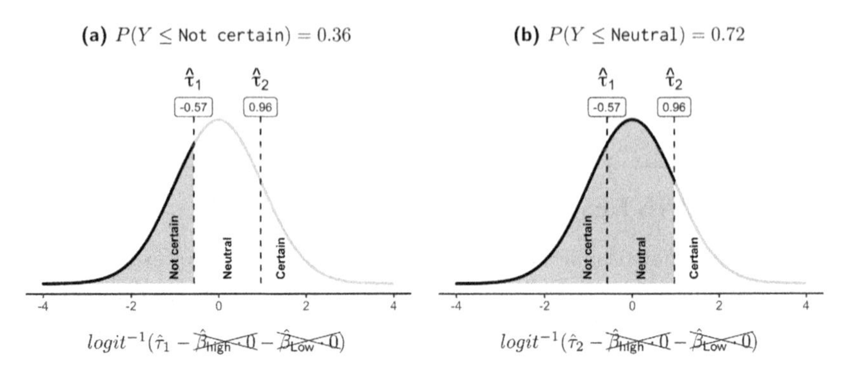

FIGURE 8.3 Calculating Probabilities in Ordinal Models (Condition = NoBreak)

therefore cumulative because it includes Neutral and all the categories *below* Neutral, that is, Not certain. As you can see in Fig. 8.3b, the probability we get by taking the inverse logit of $\hat{\tau}_2$ is $P = 0.72$ (again: this is all for condition NoBreak).

Given what we now know, what's the probability of choosing Certain for the NoBreak condition? Because probabilities add up to 1, the entire area under the curve in Fig. 8.3 is 1. Therefore, if we simply subtract 0.72 from 1, we will get the probability of choosing Certain (assuming the NoBreak condition): $1-0.72 = 0.28$ (note that this is *non*-cumulative). How about the non-cumulative probability of choosing Neutral? We know that the cumulative probability of choosing Neutral is 0.72, and we also know that the cumulative probability of choosing Not certain is 0.36. Therefore, the probability of choosing Neutral must be $0.72 - 0.36 = 0.36$. Look at Fig. 8.2 and check whether these probabilities make sense.

In summary, by taking the inverse logit of the $\hat{\tau}$ values in our model, we get predicted cumulative probabilities for condition NoBreak, our reference level here. If we wanted to calculate cumulative probabilities for conditions High and Low, we would have to plug in 1s and 0s to the equation, and our figures would look different since the areas under the curve would depend not just on $\hat{\tau}_{1,2}$ but also on the estimates of our predictors. The take-home message is that $\hat{\tau}$ values depend on where we are on the scale (which category we wish to calculate cumulative probabilities for). In contrast, $\hat{\beta}$ estimates are *the same* for each and every category in the scale for which we calculate cumulative probabilities (taking the inverse logit of the log-odds predicted by the model).

Let's calculate by hand the cumulative probabilities for condition Low using the estimates in fit_clm1. In 8.2, we see the cumulative probability for the first category j in our scale (Not certain), which is 0.58. In 8.3, we see the cumulative probability for the second category j (Neutral), which is 0.86.

Remember: that is the probability of choosing Neutral *or less*. As a result, we can now calculate the non-cumulative probability of choosing Neutral: 0.86 − 0.58 = 0.28.

$$
\begin{aligned}
P(J \leq j = 1) &= logit^{-1}(\hat{\tau}_1 - \hat{\beta}_{high} x_{i_{high}} - \hat{\beta}_{low} x_{i_{low}}) \\
P(J \leq j = 1) &= logit^{-1}(-0.57 - \xcancel{1.08 \cdot 0} - (-0.88 \cdot 1)) \\
P(J \leq j = 1) &= logit^{-1}(0.31) = \boxed{0.58} \quad \text{1em (Not certain (or less))}
\end{aligned} \tag{8.2}
$$

$$
\begin{aligned}
P(J \leq j = 2) &= logit^{-1}(\hat{\tau}_2 - \hat{\beta}_{high} x_{i_{high}} - \hat{\beta}_{low} x_{i_{low}}) \\
P(J \leq j = 2) &= logit^{-1}(0.96 - \xcancel{1.08 \cdot 0} - (-0.88 \cdot 1)) \\
P(J \leq j = 2) &= logit^{-1}(1.84) = \boxed{0.86} \quad \text{(Neutral or less)}
\end{aligned} \tag{8.3}
$$

Finally, because we know that we only have three categories (J = 3), we can now calculate the non-cumulative probability of choosing Certain: 1 − 0.86 = 0.14. As you can see, we can calculate different individual (non-cumulative) probabilities for any level j once we have the cumulative probabilities of J − 1 levels.

$$
P(j = 1) = 1 - 0.86 = \boxed{0.14} \quad \text{(Certain)} \tag{8.4}
$$

Naturally, we won't calculate anything by hand, but I hope the earlier calculations helped you see how an ordinal model works (intuitively). Instead, we can simply use the predict() function once more. Code block 44 shows how you can generate all predicted probabilities for all three conditions for all three categories in our scale—notice that we are generating non-cumulative probabilities (type = "prob" in line 6). If you look at line 19, you will see all three probabilities manually calculated earlier: 0.58, 0.28, and 0.14.[5] You can also see that if you add the first two columns for condition Low, you will get the cumulative probability calculated earlier as well.

Finally, make sure you go back to Fig. 8.2 to compare the actual proportions in the data to the predicted probabilities of the model. They should look very similar, since we're predicting the same dataset used in our model. We will now turn to a more complex model,which contains an interaction between Condition and L1.

8.2.2 *Do* Condition *and* L1 *Interact?*

Thus far, we have modeled only part of our data, namely, only data points from Spanish speakers. As a result, our first model run earlier (fit_clm1) is *not* aligned with Fig. 8.2 (cf. §2.6.2.1). Most of the time, you shouldn't do that: our figures should mirror what our models are analyzing. There are, of course, exceptions

R code

```
# Generate predicted probabilities:
predictedProbs1 = as_tibble(predict(fit_clm1,
                            newdata = tibble(Condition = c("High",
                                                           "Low",
                                                           "NoBreak")),
                            type = "prob")$fit)

# Add column for Condition:
pred_fit_clm1 = predictedProbs1 %>%
  mutate(Condition = c("High", "Low", "NoBreak"))

# Print tibble:
pred_fit_clm1

# A tibble: 3 x 4
#   `Not certain` Neutral Certain Condition
#           <dbl>   <dbl>   <dbl> <chr>
# 1         0.161   0.310   0.529 High
# 2         0.576   0.287   0.137 Low
# 3         0.361   0.363   0.276 NoBreak
```

CODE BLOCK 44 Predicted Probabilities for Ordinal Model

to this rule. For example, we could argue that learners and native speakers have fundamentally different grammars. And if we want to equate a statistical model to a grammar, then we could have one model per grammar, that is, one model for learners and one model for native speakers. Naturally, doing that will make it harder to compare the two groups. The bottom line is that your approach will depend on your goals and on your research questions.

If you look back at Fig. 8.2, you will know the answer to the question "Do Condition and L1 interact?" The answer is clearly "yes", given the trends in the figure: if we only consider the Certain category, English speakers (controls) are more certain in the Low condition (50%) than Spanish speakers are (14.5%). In our next model, which we will call fit_clm2, we will include Condition and L1, as well as their interaction as predictors.

To interpret the results from fit_clm2 in code block 45, it will be useful to go back to Fig. 8.2 once more. Our model has five coefficients ($\hat{\beta}$)—and two cutpoints ($\hat{\tau}$). Thus, we have a total of seven parameters. The only non-significant estimate is that of L1Spanish ($\hat{\beta} = -0.17, p = 0.46$). Before we actually interpret the results, notice that we see Spanish, but not English, and we see High and Low, but not NoBreak. This should already clarify what the reference levels are (i.e., the levels we don't see in the output)—and should be familiar from previous chapters.

What does L1Spanish represent here? In a nutshell, it represents the effect of being a Spanish speaker (*vs.* an English speaker) for the condition NoBreak. Look back at Fig. 8.2 and focus on the NoBreak condition in the plot. You will notice that both groups behave similarly as far as their certainty levels

R code
```
fit_clm2 = clm(Certainty3 ~ Condition * L1, data = rc, link = "logit")
summary(fit_clm2)

# Coefficients:
#                         Estimate Std. Error z value Pr(>|z|)
#   ConditionHigh          -0.6426     0.2644  -2.430  0.01508 *
#   ConditionLow            0.7430     0.2697   2.755  0.00587 **
#   L1Spanish              -0.1696     0.2302  -0.737  0.46133
#   ConditionHigh:L1Spanish 1.7096     0.3262   5.241 1.60e-07 ***
#   ConditionLow:L1Spanish -1.6126     0.3316  -4.864 1.15e-06 ***
#   ---
#   Signif. codes:  0 '***' 0.001 '**' 0.01 '*' 0.05 '.' 0.1 ' ' 1
#
# Threshold coefficients:
#                           Estimate Std. Error z value
# Not certain|Neutral        -0.7263     0.1941  -3.742
# Neutral|Certain             0.7754     0.1944   3.989
```

CODE BLOCK 45 Modeling Certainty As a Function of Condition and L1

go. In other words, only by looking at the High and Low conditions do we see more substantial differences between the two groups of participants. So it's not surprising that we don't have a significant coefficient for L1Spanish.

The coefficient for L1Spanish is negative (the null hypothesis, as usual, is that $\hat{\beta} = 0$). A negative estimate indicates that the predictor in question increases the probability of a response on the lower end of the scale (i.e., Not certain). Look again at Fig. 8.2. You will notice that Spanish speakers have 27.5% of Certain in their responses, while English speakers have 36% of Certain in their responses. We can see that Spanish speakers are therefore leaning more towards the lower end of the scale relative to English speakers (for the NoBreak condition)—hence the negative estimates. However, as we see in the output of fit_clm2, that difference is not statistically significant—so we fail to reject the null hypothesis that $\hat{\beta} = 0$ (log-odds).

Let's now spend some time interpreting the estimates that are significant in our model. Remember: by looking at the sign of the coefficients, we can see whether a predictor moves participants' certainty levels left or right along our scale. But to calculate exact probabilities we need to include the $\hat{\tau}$ values estimated in our model. First, let's focus on the effect sizes. Later we will plot the exact percentages in a clear and intuitive way.

CONDITIONHIGH. This is the effect (in log-odds) of High relative to our baseline, that is, NoBreak, for English speakers only. The effect is negative, so when we have high attachment stimuli (*vs.* stimuli with no breaks), participants' certainty levels go down (i.e., they move towards the lower end of the scale). Again: look back at Fig. 8.2 and focus on the L1: English facet only.

You will notice that for the NoBreak condition, we have 36% of Certain responses. Now look at the High condition: we have 15% of Certain. Certainty is going down in the High condition, and that's why our estimate is negative here.

CONDITIONLOW. Here we see the opposite effect: a positive estimate. Looking at Fig. 8.2, this again makes sense: while for No break we have 36% of Certain, for Low we have 50%. So in this case certainty is going up relative to NoBreak. Therefore, our estimate is *positive*. Here, the Low condition increases English speakers' certainty levels by a factor of 2.10 ($e^{|0.743|}$)—note that we are talking about the scale as a whole, so we can't be precise about which categories of the scale we are referring to.[6]

CONDITIONHIGH:L1SPANISH. Interactions are always trickier to interpret (you may remember this from the model in code block 39 back in chapter 7). Figures are especially important to visualize interactions and understand what interaction terms mean in a model. So let's look back at Fig. 8.2, this time focusing on the High condition at the bottom of the figure. For English speakers, we have 45% of Not certain responses, and 15% of Certain. For Spanish speakers, we have only 15.5% of Not certain responses and 52.5% of Certain. Clearly, these two language groups differ when it comes to the High condition: Spanish speakers are much more certain—which makes sense, given that Spanish favors high attachment. Likewise, if you only look at Spanish speakers and compare High and NoBreak, you will again see that there's a big difference in percentage between the two conditions. Simply put, this is what the interaction is telling us: the effect of Condition depends in part on whether you are an English speaker or a Spanish speaker in our hypothetical study. Notice that the estimate of our interaction here is positive ($\hat{\beta} = 1.71, p < 0.001$). This makes sense: after all, the High condition (*vs.* NoBreak) makes Spanish (*vs.* English) speakers much more certain. Another way of looking at it is to visually compare NoBreak and High across both groups in Fig 8.2. You will notice that they follow opposite patterns—hence the significant interaction in our model.

CONDITIONLOW:L1SPANISH. The interpretation of this estimate should now be clear. Here we have a negative effect ($\hat{\beta} = -1.61, p < 0.001$), which captures the observation in Fig. 8.2 that Spanish speakers are *less* certain than English speakers in the Low condition (*vs.* in NoBreak stimuli).

What do the cut-points tell us? Remember: like any intercept, the two $\hat{\tau}$ values here assume that all predictors are set to zero (i.e., Condition = NoBreak and L1 = English). If you take the inverse logit of these two values, you will get cumulative probabilities of being Not certain (or less) and Neutral (or less)—the same meaning discussed for fit_clm1 earlier. If you run invlogit() on $\hat{\tau}_1$, you will get 0.33; if you run it on $\hat{\tau}_2$, you will get 0.68. Therefore, there's a 68% probability that English speakers in the

NoBreak condition will be Neutral or less about their responses. Here, again, if you want to calculate the actual probabilities for Spanish speakers or for other conditions in the study, you will need to go through the estimates like we did for fit_clm1. We won't do that here because it's much easier (and faster!) to simply use the predict() function.

We can see here that the interpretation of interactions is the same in ordinal regressions as it is in linear or logistic regressions. That makes sense: these are all generalized linear models, which means they belong to the same family of models. In other words, if you know how to interpret an interaction, you will know how to do it across all the models covered in this book (and many others in the same family, such as Poisson regressions).

Our final step in this chapter is to plot our model's predictions. What are the predicted probabilities for each category on our scale for each group for each condition? In a way, we want to reproduce Fig. 8.2, only this time instead of percentage points (proportions), we will have predicted probabilities coming from our model (fit_clm2). Remember: these will likely be similar, because we are basically testing our model with the exact same dataset it used to "study" our variables.

We will not plot the model's estimates—but you will practice doing that at the end of this chapter. Instead, we first calculate predicted probabilities (non-cumulative) using predict(). Then we create a new tibble to store the values, and use that to create a figure using ggplot2. The figure we want to plot is Fig. 8.4.

If you compare Fig. 8.4 with Fig. 8.2, you will notice they are very similar, even though they are showing two different things (predicted probabilities *vs.* actual values in our data). The code to generate Fig. 8.4 is shown in code block 46 (which you should add to plotsCatModels.R).[7] The figure itself has a series of aesthetic adjustments, which you can study in lines 23–37—play around with the code so you can see what each line is doing. Most of the code should be fairly familiar at this point, and you should also be able to adapt it to your needs (should you wish to plot predicted probabilities in the future).

8.3 Summary

In this chapter, we examined how to model scalar (or ordinal) response variables. The model we used is essentially a logistic regression, but there are some important differences (the main difference is that we now deal with more than one intercept)—Agresti (2010) offers comprehensive coverage of ordinal models. Following are some of the key points discussed earlier. Remember: a lot of what we discussed in previous chapters is also applicable to ordinal regressions. For example, we didn't rescale our variables in fit_clm2 (since they were all binary in our model), but you may want to

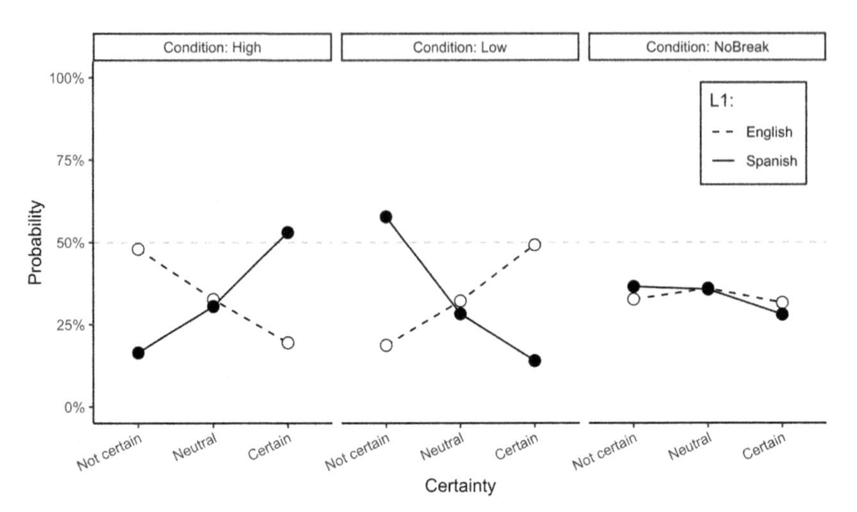

FIGURE 8.4 Predicted Probabilities Using ggplot2

do that if you're running an ordinal model with multiple variables which use different units. Finally, being able to see the data patterns in a figure can certainly help us understand our model and its output—even if you later decide not to include the figure in your paper, chances are it will improve your understanding of the results and how they relate to the data being modeled.

- In an ordinal model, we predict an ordinal (scalar) variable with J categories. In this chapter, our scale had three categories, so $J = 3$ (Not certain, Neutral, and Certain).
- Although each category in our scale has its own $\hat{\tau}$, the $\hat{\beta}$ coefficients in our models are the same across *all* categories on the scale.
- Ordinal models generate multiple intercepts, which we refer to as cutpoints, thresholds, or τ cuts. The number of $\hat{\tau}$ values estimated by our model is $J - 1$—that's why we dealt with two $\hat{\tau}$ values in fit_clm1 and fit_clm2.
- Like logistic regressions, the ordinal models we examined in this chapter (ordered logit models) also give us estimates in log-odds. As a result, we can use the invlogit() function from the arm package to get probabilities—only now these probabilities are cumulative.
- We can interpret our $\hat{\tau}$ values much like we'd interpret a typical intercept, that is, they will give us the cumulative probability of falling in category j assuming all other coefficients are zero.
- The sign of our estimates indicate whether a given predictor increases the probability of a response in the lower end of the scale (negative sign) or in the higher end of the scale (positive sign).

R code
```
# Generate predicted probabilities:
predictedProbs2 = as_tibble(predict(fit_clm2,
                                    newdata = tibble(Condition = rep(c("NoBreak",
                                                                       "Low",
                                                                       "High"),
                                                                     times = 2),
                                                     L1 = rep(c("English", "Spanish"),
                                                              each = 3)),
                                    type = "prob")$fit)

# Add L1 and Condition columns to predictions:
pred_fit_clm2 = predictedProbs2 %>%
  mutate(L1 = rep(c("English", "Spanish"), each = 3),
         Condition = rep(c("NoBreak", "Low", "High"), times = 2))

# Wide-to-long transform tibble:
longPredictions = pred_fit_clm2 %>%
  pivot_longer(names_to = "Certainty",
               values_to = "Probability",
               cols = `Not certain`:Certain)

# Plot model's predictions (load scales package first):
ggplot(data = longPredictions, aes(x = Certainty, y = Probability)) +
  geom_hline(yintercept = 0.5, linetype = "dashed", color = "gray80") +
  geom_line(aes(group = L1, linetype = L1)) +
  geom_point(aes(fill = L1), shape = 21, size = 3) +
  facet_grid(~Condition, labeller = "label_both") +
  coord_cartesian(ylim = c(0,1)) +
  scale_y_continuous(labels = percent_format()) +
  scale_x_discrete(limits = c("Not certain", "Neutral", "Certain")) +
  scale_fill_manual(values = c("white", "black"), guide = FALSE) +
  scale_linetype_manual(values = c("dashed", "solid")) +
  theme_classic() +
  labs(linetype = "L1:") +
  theme(legend.position = c(0.9,0.8),
        legend.background = element_rect(color = "black"),
        axis.text.x = element_text(angle = 25, hjust = 1))

# ggsave(file = "figures/ord-predictions.jpg", width = 7, height = 4, dpi = 1000)
```

CODE BLOCK 46 How to Plot Predicted Probabilities for Ordinal Model

- Just like linear and logistic regressions, we can use the predict() function to check the intuition behind the output of the model and to plot predicted probabilities for new (or existing) data.
- The code in Appendix F demonstrates how to run two simultaneous ordinal models, one for each language group—you can adapt the code to run linear and logistic models as well.

8.4 Exercises

Ex. 8.1 Figures and Models

1. Modify code block 42, which generates Fig. 8.2, to include only learners. You should do that inside ggplot(), and you should use the filter() function.

The resulting figure will be better aligned with fit_clm1. Finally, change the font family to Times. You will need to adjust two layers of the figure. This is because geom_text() is independent and requires its own font specification if you want the percentages inside the bars to use Times as well—you should add family = "Times" to geom_text(). To change the font family for everything else in the figure, check chapter 5.

2. In this chapter, we ran our models on a 3-point scale, but our dataset also has a 6-point scale (Certainty column), which is its raw/original ordinal response variable. By doing this, we simplified or compressed our scale (from 6 to 3 points). Run a new model, fit_clm3, which will be equivalent to fit_clm2. This time, however, use Certainty as the response variable. How do the results compare (estimates and *p*-values) to those in fit_clm3?
3. Create a figure showing the estimates for fit_clm2 using the sjPlot package.

Notes

1. Another option would be to use probit models, which we won't discuss in this book.
2. See also the polr() function from the MASS package (Venables and Ripley 2002).
3. Appendix F demonstrates how to run two simultaneous models, one for each language group, and print a table comparing estimates from both models.
4. So while a negative sign indicates a shift to the left on our scale (Not certain), a positive sign indicates a shift to the right (Certain).
5. Discrepancies are due to rounding in the manual calculation.
6. This is usually what we want. Most of the time our questions focus on whether we go up or down our scale as a whole; that's why you normally don't need to interpret the $\hat{\tau}$ values estimated.
7. We have already loaded the scales package in plotsCatModels.R, which you will need to run the code.

9

HIERARCHICAL MODELS

9.1 Introduction

So far we have examined three types of models: linear, logistic, and ordinal regressions—all of which are part of the same family of models (generalized linear models). In chapter 6, for example, we predicted the score of our participants on the basis of two variables, namely, Hours and Feedback. One important problem of all the models we ran in chapters 6, 7, and 8 is that they all assume that data points are *independent*. In reality, however, data points are rarely independent.

Take our longFeedback dataset (created in dataPrepLinearModels.R). There, each participant either belonged to the Recast group or to the Explicit correction group. Each participant completed two tasks, and each task had five assessments throughout the semester. As a result, every single participant in our hypothetical study completed ten assignments in total. The scores for these ten assignments are clearly *not* independent, since they all come from the same participant. Likewise, the five assignments from each task are not independent either, since all five come from the same task—this grouping is illustrated in Fig. 9.1. We could deal with this lack of independence by averaging the scores for all five assignments given earlier (per task) or by averaging all ten assignments, such that each participant would have one (mean) score. The issue is that every time we take the average of different values we lose important information.

In short, our models so far completely ignore that our observations come from a grouped structure, for example, Fig. 9.1. Consequently, they are blind to the fact that participants (or groups of participants) behave differently or that experimental items may elicit different types of responses. You can see

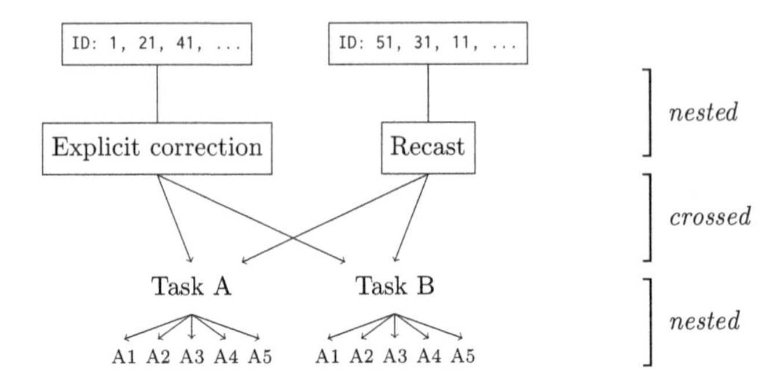

FIGURE 9.1 Grouped Data in longFeedback

why that's a major issue in studies with human subjects such as those we carry out in second language research.

Before we continue, it's essential to understand what a "grouped structure" means, which brings us to the difference between *nested* and *crossed* (i.e., non-nested) factors. Observe Fig. 9.1, which illustrates the design of our hypothetical study on the effects of feedback on learning. Here, both Explicit correction and Recast have two tasks, namely, tasks A and B. Task A is the same regardless of which feedback group we consider. Note that the arrows from the feedback groups to the tasks are crossing. In this design, Task is *crossed* with Feedback: participants in both feedback groups will see the same tasks.

Now look at the participants' IDs at the top and the tasks at the bottom of Fig. 9.1. Each task has five assignments (the Item column in longFeedback), A1 to A5. But the assignments from task A are different from the assignments from task B (even though they have the same label across tasks). Therefore, Item is *nested* inside Task. Likewise, our participant variable, ID, is nested inside Feedback: a given participant could only be part of *one* feedback group. As we consider grouped factors and how to account for them in our models, it's extremely important to fully understand not only our data but also the study design in question.

Take participant whose ID is Learner_21, a 20-year-old male (see longFeedback). This participant is in the Explicit correction group, as we can see in Fig. 9.1, so by definition he is not in the Recast group, given the nested design between Feedback and ID. However, he completed both tasks and all their respective assignments just like any other participant in the Recast group, given the crossed design between Feedback and Task.

To deal with grouped data, we could employ repeated measures analysis of variance (ANOVAs), which are very commonly used in the literature. In this book, however, we won't do that. Instead, we will explore a more powerful method, namely, *hierarchical models*, also known as multilevel models. Unlike repeated measures ANOVAs, these models allow for multiple grouping variables, and they can also handle unbalanced data quite well.

Hierarchical models combine fixed effects with random effects—for that reason, these models are also known as *mixed-effects models*, or simply *mixed models*. Fixed effects are the effects we are interested in—the same effects we examined in previous chapters. For the longFeedback data, for example, Feedback, Task, and Hours would be fixed effects. Random effects are effects we are not directly interested in but that must be accounted for given that there are groups of observations that we don't expect to be independent, that is, that violate the independence assumption briefly discussed earlier.

Consider the variable ID, which identifies each participant in longFeedback. Participants' scores are not independent (i.e., each participant has ten scores in the data), so that must be accounted for. At the same time, we don't want to model individual behavior—our participants were randomly sampled from the population, and if we were to replicate our study, participants would be different again. Instead, we want to generalize the overall patterns of our fixed effects beyond the random individual variation we observe in the data. Finally, we'll assume in this chapter that we are only interested in fixed effects—effects that generalize across participants and items (and which, in previous chapters, are referred to simply as *effects*).

So let's see how hierarchical models work. Any model examined in the previous chapters can be made hierarchical, but we will use linear regressions here. In 9.1, we see a typical linear regression with one predictor variable, x. Note that the intercept, β_0, is assumed to be the same for *all* participants. This is clearly not ideal, given how much people vary in their responses—humans are not known to be consistent, after all. Likewise, the effect of our predictor variable, β_1 (i.e., our slope), is also assumed to be the same for all participants. In other words, not only do we assume that everyone in the data has the same intercept, we also assume that everyone is affected by our predictor variable the same way.

$$y_i = \beta_0 + \beta_1 x_{i1} + \epsilon_i \tag{9.1}$$

The model in 9.1 ignores the complex structure of our data and the variation coming from different participants, items, and so on. Such a model can give us estimates for a given predictor $\hat{\beta}$ that may be incorrect once we take into account how much participants and items vary in our data. In particular, Type I or Type II errors become more likely: we could reject the null hypothesis when it's in fact true (Type I error), or we could fail to reject the null hypothesis when it's in fact false (Type II error).

If we assume that the model in question is defined as Score ~ Task, then our intercept represents the average score for task A, and our estimate represents the slope from task A to task B (i.e., the difference between the two tasks). This model fit is shown in Fig. 9.2a—all three plots in Fig. 9.2 exhibit the empirical patterns in longFeedback. We can improve our model

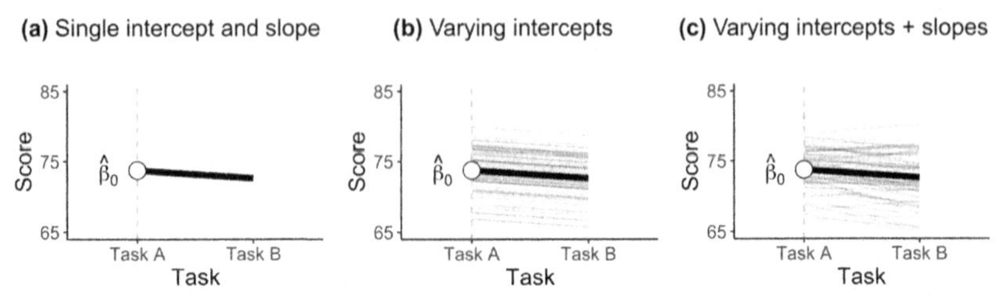

FIGURE 9.2 Simple Model (a) *vs.* Hierarchical Models (b–c)

by adding random intercepts (e.g., by participant), shown in 9.2—which gives us a simple hierarchical model, illustrated in Fig. 9.2b.

$$y_i = \beta_0 + \boxed{\alpha_{j[i]}} + \beta_1 x_{i1} + \epsilon_i \qquad (9.2)$$

In 9.2, we have a new term, $\alpha_{j[i]}$, which works as our by-participant intercept. Here, j represents ID and $j[i]$ represents the ith participant. Thus, $\alpha_{j[i]}$ represents how far from β_0 each participant is—that is, it is the offset of each participant relative to β_0. In the visual representation of this type of model shown in Fig. 9.2b, every participant has an individual intercept, but the slope of the all the lines is the same, so all the lines are parallel (i.e., we assume that the effect of task B is constant across participants).

Finally, we can also make our slopes vary by participant. This is shown in 9.3, where we have one offset for the intercept and one offset for our predictor estimate, $\gamma_{j[i]}$. As we can see in Fig. 9.2c, participants seem to vary in both intercept and slope; both kinds of variation are accounted for in model 9.3.

$$y_i = \beta_0 + \boxed{\alpha_{j[i]}} + (\beta_1 + \boxed{\gamma_{j[i]}})x_{i1} + \epsilon_i \qquad (9.3)$$

We will see how to run hierarchical models in the next section, as usual, but for now let's spend some more time going over Fig. 9.2. As already mentioned, when we run a model, we are mostly interested in our fixed effects—here, Task is a fixed effect, while our varying intercept and slope are our random effects. However, individual variation is a key characteristic of data in second language studies. This is yet another reason that we should favor hierarchical models over simple (non-hierarchical) models.

Did you notice that the solid thick line in Fig. 9.2 is essentially the same across all three figures? In other words, our fixed effects here, $\hat{\beta}_0$ and $\hat{\beta}_1$, are basically the same whether or not we add random effects to our model. This is because the data in longFeedback is balanced, and no one participant behaves too differently relative to other participants—which is often *not* the case in second language research. If we had unbalanced data, or individual participants whose behavior were too different from other participants, the estimates from our hierarchical models in Fig. 9.2 would look different compared to our non-hierarchical model.

While fixed effects might not change in a hierarchical model, the standard errors estimated for these fixed effects almost always look different from those in a non-hierarchical model. Intuitively, this makes sense: the more we take into account the variation in our data, the more precise our estimates will be. Precision can mean that our standard errors will be larger, which will reduce Type I errors; it can also mean that our standard errors will be smaller, which will reduce Type II errors (see, for example, discussion on fit_lmer1 later). In general, we are more worried about the former.

Before we see hierarchical models in action, it's important to understand the structure of our dataset here. Notice that Fig. 9.2c has by-participant random intercepts and slopes (for Task). That structure is possible because every participant completed the same tasks (crossed design in Fig. 9.1). If we had a nested design for tasks, we wouldn't be able to run such a model. For example, our Feedback variable cannot be specified as a random effect by participant, because no participant actually tried both feedback styles! As you can see, to think about random effects we have to understand the design of our study and the structure of our data.

Finally, Table 9.1 lists three different model specifications in R. Model A is a typical linear model predicting participants' scores as a function of Task—this should be quite familiar at this point. You can think of model A as our "baseline", since it's a simple non-hierarchical model. Model B includes a varying intercept for each participant and a varying intercept for each item (assignment)—"1" there simply means "intercept". This model takes into account the fact that we expect to see variation across participants and also across assignments in our study. Finally, model C includes a varying slope for Task—this is essentially what you see in Fig. 9.2c, where both intercepts and slopes are allowed to vary by participant. The most important difference between models B and C is that model C assumes that the effect of task will be different for different participants—which is a very reasonable assumption.

The random slope that we see in model C in Table 9.1 reflects the crossed structure of our data, where every participant took part in both tasks. Again: we wouldn't be able to add a by-participant random slope for Feedback, given that participants were never in both feedback groups in our hypothetical study—we

TABLE 9.1 Model Specification in R Using Task as a Predictor

Model	*Code in R*
A.	Score ~ Task *baseline: non-hierarchical*
B.	Score ~ Task + (1 | ID) + (1 | Item)
C.	Score ~ Task + (1 + Task | ID) + (1 | Item)

could, however, add a by-item (assignment) random slope for Feedback (see Fig. 9.1). We will discuss model structures in more detail in the next section, where we will run different hierarchical models.

Finally, notice that all the models we have discussed so far can be easily adapted to be hierarchical. The syntax in Table 9.1 works consistently for all the linear, logistic, and ordinal regressions discussed in this book: you can run hierarchical logistic regressions using the glmer() function in the lme4 package[1] and hierarchical ordinal regressions using the clmm() function in the ordinal package. Running hierarchical versions of our models in R is actually quite easy, i.e., the code is nearly identical to what we have already seen. The tricky part is not the code, but the statistical concept behind it, and which random effects to specify in our model given the data.

9.2 Examples and Interpretation

In this section, we will run two hierarchical models—and a simple, non-hierarchical model, so we can compare our results. First, let's create one more script for our hierarchical models: hierModels.R. This script should go inside the Frequentist folder, which already contains the data file we'll be using, feedbackData.csv. Before we get started, you will need to install the lme4 package (Bates et al. 2015b), which is the most common package used for mixed effects in linguistics. At the top of hierModels.R, you should source dataPrepLinearModels.R and load both lme4 and arm. Once you source dataPrepLinearModels.R, you will automatically load the feedback and longFeedback tibbles. This time, however, we will improve on our dataPrepLinearModels.R code—recall that this was originally presented in code block 11 back in chapter 3. So replace the simpler code you have in the dataPrepLinearModels.R with the code in code block 47.

R code
```
1  library(tidyverse)
2
3  # Read file as tibble:
4  feedback = read_csv("feedbackData.csv")
5
6  # Wide-to-long transform for models:
7  longFeedback = feedback %>%
8    pivot_longer(names_to = "Task",
9                 values_to = "Score",
10                cols = task_A1:task_B5) %>%
11   separate(col = Task, into = c("Task", "Item"), sep = 6) %>%
12   mutate(Task = ifelse(Task == "task_A", "Task A", "Task B")) %>%
13   mutate_if(is.character, as.factor)
```

CODE BLOCK 47 Updated Content of dataPrepLinearModels.R

In code block 47, line 12 renames the levels of Task, and line 13 simply transforms all character variables into factor variables (this was already in code block 11, line 9). Notice that the code also loads tidyverse (line 1), as usual. Now we're ready to analyze our data.

9.2.1 Random-Intercept Model

Let's run a model with both Feedback and Task as predictors. Our response variable will be Score—we will use the longFeedback tibble. As usual, let's take a look at a figure that plots the variables we'll be modeling later. Fig. 9.3 clearly shows that (i) the score for Recast is higher than the score for Explicit correction, (ii) participants had higher scores for Task A than for Task B, and (iii) these two variables don't seem to interact (the lines are parallel). The mean score per feedback group (ignoring tasks) is represented by "**x**" in the figure. The code that generated Fig. 9.3 is shown in code block 48 and should go in plotsLinearModels.R (see Appendix D), since our hierarchical model here is a linear regression.

Now, let's run our model. First, let's run a simple non-hierarchical linear regression, fit_lm0. Then let's compare it with a hierarchical linear regression with by-participant and by-item random intercepts. Both models are run in code block 49, and you see their respective outputs printed as comments in

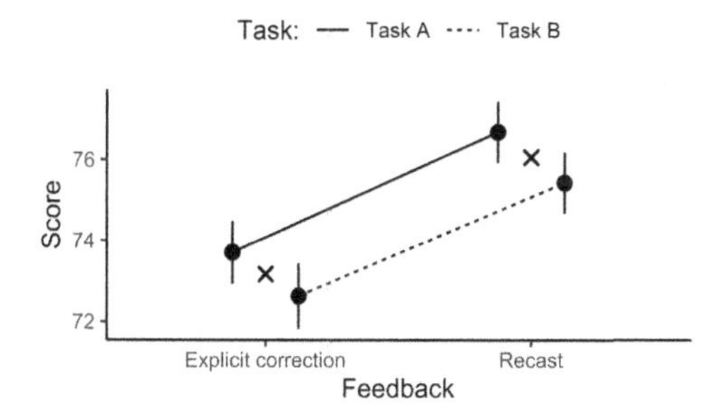

FIGURE 9.3 Effect of Feedback and Task on Participants' Scores

R code
```
source("dataPrepLinearModels.R")

ggplot(data = longFeedback, aes(x = Feedback, y = Score)) +
  stat_summary(aes(group = Task), position = position_dodge(width = 0.5)) +
  stat_summary(fun = mean, shape = 4) +
  stat_summary(aes(group = Task, linetype = Task),
               geom = "line", position = position_dodge(width = 0.5)) +
  theme_classic() +
  theme(legend.position = "top") +
  labs(linetype = "Task:")

# ggsave(file = "figures/scores-feedback-task.jpg", width = 4, height = 2.5, dpi = 1000)
```

CODE BLOCK 48 Plotting Scores by Feedback and Task

R code

```
source("dataPrepLinearModels.R")
library(lme4)
library(arm)
library(MuMIn) # To calculate R-Squared for hierarchical models

fit_lm0 = lm(Score ~ Feedback + Task, data = longFeedback)
fit_lmer1 = lmer(Score ~ Feedback + Task + (1 | ID) + (1 | Item), data = longFeedback)

display(fit_lm0)

# lm(formula = Score ~ Feedback + Task, data = longFeedback)
#                coef.est  coef.se
# (Intercept)     73.75     0.67
# FeedbackRecast  2.89      0.77
# TaskTask B     -1.17      0.77
# ---
# n = 600, k = 3
# residual sd = 9.47, R-Squared = 0.03

display(fit_lmer1)

# lmer(formula = Score ~ Feedback + Task + (1 | ID) + (1 | Item),
#      data = longFeedback)
#                coef.est  coef.se
# (Intercept)     73.75     3.58
# FeedbackRecast  2.89      1.01
# TaskTask B     -1.17      0.44
#
# Error terms:
# Groups    Name        Std.Dev.
# ID        (Intercept) 3.51
# Item      (Intercept) 7.83
# Residual              5.33
# ---
# number of obs: 600, groups: ID, 60; Item, 5
# AIC = 3840.4, DIC = 3841
# deviance = 3834.7
```

CODE BLOCK 49 Running a Model with Random Intercepts: Feedback and Task

the same code block. Because both outputs are printed using display(), not summary(), you won't see *p*-values. Later I discuss why we will now stop looking at *p*-values. You should add code block 49 to hierModels.R.

First, in lines 11–18 we see the output for fit_lm0, a non-hierarchical linear regression. We have two columns: coef.est (the estimates of our predictors) and coef.se (their standard errors). We could calculate confidence intervals with these two columns, and if our interval includes zero, that would be equivalent to a non-significant *p*-value (§1.3.4). Alternatively, we can simply run confint(fit_lm0)—recall that these two methods are *not* identical (chapter 6). FeedbackRecast also has a reliable positive estimate, but let's focus on TaskTask B. There, the effect is −1.17 and the standard error is 0.77.

fit_lmer1 adds a random intercept by participant, (1 | ID), and a random intercept by item, (1 | Item)—so this model has two α terms as well as two β terms (and the intercept, β_0). Note that now our output for fit_lmer1 comes in two parts: the first part can be found in lines 24–27, our fixed effects—this is what we care about. Then we have the error terms (or random effects) in lines 29–33. Let's focus on the fixed effects first.

The fixed effects for fit_lmer1 are the same as those we discussed earlier for fit_lm0—indeed, we interpret the model estimates the same way we interpret our estimates back in chapter 6. However, pay attention to the standard errors, as they're very different. First, the standard error for our intercept went from 0.67 to 3.58, and the standard error for FeedbackRecast went from 0.77 to 1.01. Second, the standard error for our task effect went from 0.77 in fit_lm0 to 0.44 in fit_lmer1. Because the effect sizes of our intercept and FeedbackRecast are large relative to their SEs, the increase in their SE is not sufficient to invalidate their statistical effect (i.e., $t > 1.96$). On the other hand, by accounting for the variation coming from participants and items, we now have a statistically significant effect of Task. If you calculate the 95% confidence interval for TaskTask B here, you will notice that it no longer includes zero, so here we would reject the null hypothesis. Importantly, rejecting the null here makes more intuitive sense given what we see in Fig. 9.3—and is in fact the right conclusion, since the data in question was simulated and the two tasks do have different mean scores. The bottom line here is that fit_lm0 led us to a Type II error, and fit_lmer1 didn't.

Now let's inspect our error terms, shown in lines 29–33. It will be helpful now to look back at Fig. 9.2b. The thick black line represents the intercept ($\hat{\beta}_0$) and slope ($\hat{\beta}$) for Task in our model, shown in lines 25 and 27 (two of our fixed effects; Feedback is not shown in the figure). Line 31, in turn, shows us the estimated by-participant standard deviation ($\hat{\sigma} = 3.51$) from $\hat{\beta}_0$, that is, how much by-participant intercepts deviate from the intercept (grand mean; $\hat{\beta}_0 = 73.75$). Random intercepts are assumed to follow a Gaussian (normal) distribution, so we can use $\hat{\beta}_0$ and $\hat{\sigma}$ to calculate the predicted 95% interval for the participants' intercepts: $73.75 \pm (1.96 \cdot 3.51) = [66.87, 80.63]$—alternatively, simply run confint(), which will yield almost identical values here. Likewise, line 32 tells us how much scores deviated from each item (i.e., assessment within each task) relative to the mean score for all items (we don't have this information in the figure).

Another important difference between our output is the estimated standard deviation of the residual, or $\hat{\sigma}_e$ (e = error). For fit_lm0, it's 9.47 (line 18); for fit_lmer1, it's 5.33 (line 33). The intuition is simple: because our model is now including variation by participant and by item, some of the error (residual) is accounted for, and we end up with a lower $\hat{\sigma}_e$—inspect Fig. 9.2 again, and go back to the discussion on residuals in chapter 6 if you don't remember what residuals are.

At the end of code block 49, line 36 gives us AIC and DIC (deviance information criterion) values. We already discussed AIC values in chapter 7, and we will use them soon to compare different model fits.

WHERE'S R^2? You may have noticed that while the output for fit_lm0 in code block 49 prints an R^2 (line 18; R-Squared = 0.03), we don't have an R^2 value for our hierarchical model, fit_lmer1. We can generate R^2 values for hierarchical models by installing and loading the package MuMIn (Bartoń 2020) and then running r.squaredGLMM(fit_lmer1), which here will give us *two* values (cf. fit_lm0): R2m = 0.02 and R2c = 0.73 (remember that you can always run ?r.squaredGLMM to read about the function in R).[2] Here, the first value refers to our fixed effects (marginal R^2), while the second refers to the model as a whole (conditional R^2). Simply put, our fixed effects describe roughly 2% of the variance in the data, and our model describes over 70% of the variance observed.

WHERE ARE THE *P*-VALUES? We have already used display() to print model outputs before (e.g., code block 26), so this is not the first time we don't see *p*-values in an output. What's different this time, however, is that even if you try running summary(fit_lmer1), you will not get *p*-values from hierarchical models in lme4. There are ways to "force" *p*-values to be printed, of course (e.g., lmerTest package),[3] but there's a good reason to avoid them—that's why the authors of lme4 don't include them by default in our outputs. The problem here is that it's not obvious how to calculate degrees of freedom once we have a hierarchical model, so we won't bother with exact *p*-values—I want you to stop focusing on them anyway and focus instead on estimates and standard errors, which allow us to calculate confidence intervals. An easy (and better) solution is to ignore *p*-values and to calculate and report confidence intervals with the confint() function. You can read more about this topic in Luke (2017) or Kuznetsova et al. (2017), or see https://stat.ethz.ch/pipermail/r-help/2006-May/094765.html—a post by Douglas Bates, one of the authors of lme4. Soon, in chapter 10, we won't even discuss *p*-values, so now's a good time to get used to their absence. Let us now turn to our second hierarchical model—we'll see how to report our results shortly.

9.2.2 Random-Slope and Random-Intercept Model

If we only allow our intercepts to vary, we are forcing the model to assume that the effect of Task is the same for *all* participants. Clearly that's not a realistic assumption, since we normally have considerable variation in our data. In code block 50, we run a model (fit_lmer2) where both intercepts *and* slopes (for Task) are allowed to vary. Line 8 specifies our fixed effects of interest, that is, we want to use Feedback and Task to predict a participant's score. Line 9 adds a by-participant random intercept and a random slope for Task—that is, we are assuming that the effects of Task can vary depending on the participant, which is a reasonable assumption. Finally, line 10 adds a random intercept by item (something we also did for fit_lmer1 earlier). We can now draw a visual comparison between our models and Fig. 9.2: whereas fit_lmer1 mirrors Fig. 9.2b, fit_lmer2 mirrors Fig. 9.2c (and, of course, fit_lm0 mirrors Fig. 9.2a).

First, let's focus on our fixed effects. Note that the estimates are very similar, which shouldn't be surprising given what we discussed earlier. But look closely at the standard error of the estimates. For example, in fit_lmer1, the SE for TaskTask B was 0.44. Now, in fit_lmer2, it's 0.54—it went up. You can run confint(fit_lmer2) to calculate the 95% confidence intervals of our estimates, and you will see that Task here still doesn't include zero (R may take some seconds to calculate intervals).

Next, let's look at our random effects (lines 18–23). Lines 20 and 22 are already familiar from our discussion about fit_lmer1: they simply tell us how much the by-participant and by-item intercepts vary relative to their respective means. What's new here is line 21, TaskTask B. This line has two values, namely, Std.Dev. and Corr. The first number indicates how much by-participant slopes vary relative to the main slope ($\hat{\beta} = -1.17$) in our fixed effects, which represents the mean slope for Task for all participants.

Like before, we can calculate the predicted 95% interval for by-participant slopes. Here, our mean is $\hat{\beta} = -1.17$, and our standard deviation for the slope of Task is $\hat{\sigma} = 2.68$. As a result, 95% of our participants' slopes are predicted to fall between −6.42 and 4.08. Line 21 also shows the predicted correlation between the intercepts and the slopes in the model (−0.11). The negative number indicates that higher intercepts tended to have lower slopes. Finally, note that in fit_lmer1 the standard deviation of our by-participant random intercept was 3.51; in fit_lmer2, it's 3.43, so our model with a by-participant random slope has reduced the unexplained variation between participants.

How many parameters does fit_lmer2 estimate? The answer is *seven*. We have three fixed effects ($\hat{\beta}_0, \hat{\beta}_1, \hat{\beta}_2$), three random effects ($\hat{\sigma}$ for two random intercepts and one random slope), and an estimated correlation (ρ) between by-participant random intercepts for Task and by-speaker random slopes for Task.[4]

Here's an easy way to actually see our random effects. Code block 51 extracts by-participant and by-item effects from fit_lmer2. First, we create two variables (lines 2 and 3), one for participant effects (byID), and one for item effects (byItem). Let's focus on the former, since that's more relevant to us here. Lines 6–12 show you the top six rows of participant-level random effects. Notice how both columns are different for each participant—this makes sense, since we have varying intercepts and slopes in our model.

Look back at our model's estimates in code block 50. Our intercept is 73.74, and the effects of Feedback and Task are, respectively, $\hat{\beta} = 2.91$ and $\hat{\beta} = -1.17$. Now take Learner 1. This participant's intercept is −2.42, which indicates the difference between this particular speaker's intercept and

R code

```
fit_lmer2 = lmer(Score ~ Feedback + Task +
                 (1 + Task | ID) +
                 (1 | Item),
               data = longFeedback)

display(fit_lmer2)

# lmer(formula = Score ~ Feedback + Task +
#                   (1 + Task | ID) +
#                   (1 | Item),
#                 data = longFeedback)

#                coef.est  coef.se
# (Intercept)     73.74      3.58
# FeedbackRecast   2.91      0.99
# TaskTask B      -1.17      0.54
#
# Error terms:
# Groups   Name        Std.Dev.  Corr
# ID       (Intercept) 3.43
#          TaskTask B  2.68      -0.11
# Item     (Intercept) 7.83
# Residual             5.14
# ---
# number of obs: 600, groups: ID, 60; Item, 5
# AIC = 3834.9, DIC = 3832.2
# deviance = 3825.6

r.squaredGLMM(fit_lmer2)
#          R2m       R2c
# 0.02356541 0.7469306
```

CODE BLOCK 50 Running a Model with Random Intercepts and Slopes: Feedback and Task

our main intercept. The offset here is negative, which means this participant's score for task A is below average. We have a negative slope too, so this participant's slope is also below average.

Let's use Learner 1, who was in the Explicit correction group, to plug in the numbers in our model—this is shown in 9.4. It may be useful to go back to the output of fit_lmer2 in code block 50. Recall that our fixed effects are represented by $\hat{\beta}$ and our random effects are represented by $\hat{\alpha}$ (intercept) and $\hat{\gamma}$ (slope). In 9.4, $\hat{\beta}_1$ is our estimate for Feedback and $\hat{\beta}_2$ is our estimate for Task. Crucially, because our model has a by-item random intercept, we also have to be specific about which item (assignment) we want to predict scores for. Let's assume Item $=$ 3, so we'll be predicting the score for the participant's third assignment in task B. Because Feedback $=$ Explicit correction, we set x_1 to zero; and since Task $=$ Task B, we set x_2 to 1.

R code

```
1  # Check random effects for fit_lmer2:
2  byID = lme4::ranef(fit_lmer2)$ID
3  byItem = lme4::ranef(fit_lmer2)$Item
4
5  head(byID)
6  #              (Intercept) TaskTask B
7  # Learner_1     -2.420820 -1.3072026
8  # Learner_10     1.256462 -1.9007619
9  # Learner_11     0.492263 -2.5075564
10 # Learner_12     1.209074  2.5266376
11 # Learner_13    -3.898236 -3.1992087
12 # Learner_14    -2.579415  0.8593917
13
14 head(byItem)
15 #   (Intercept)
16 # 1  -9.5335424
17 # 2  -5.3627110
18 # 3  -0.1780263
19 # 4   5.2017898
20 # 5   9.8724899
```

CODE BLOCK 51 Random Effects in a Hierarchical Model (lmer)

It's worth examining 9.4 carefully to fully understand how our model works. Here's a recap of what every term means: $\hat{\beta}_0$ is our intercept, as usual; $\hat{\alpha}_{j[1]}$ is our random intercept for the participant in question (j = ID, and 1 = Learner 1); $\hat{\alpha}_{k[3]}$ is our random intercept for the item (assignment) in question (k = Item, and 3 = third item); $\hat{\beta}_1$ is Feedback = Recast; $\hat{\beta}_2$ is Task = Task B; and, finally, $\hat{\gamma}_{j[1]}$ is the random slope for Task for Learner 1. All the numbers we need are shown in code blocks 50 and 51.

The predicted score for the conditions in question is 68.67, which is not too far from the participant's actual score, 68.2—you can double-check this entire calculation by running code block 52. The mean score for the explicit correction group is 73.2—represented by "x" in Fig. 9.3. Therefore, the predicted score for Learner 1 is a little closer to the group mean relative to the participant's actual score.

$$\begin{aligned}
\hat{y}_1 &= \hat{\beta}_0 + \hat{\alpha}_{j[1]} + \hat{\alpha}_{k[3]} + \hat{\beta}_1 x_{1_1} + (\hat{\beta}_2 + \hat{\gamma}_{j[1]}) x_{2_1} + \hat{e}_1 \\
\hat{y}_1 &= 73.74 + (-2.42) + (-0.17) + \cancel{2.91 \cdot 0} + (-1.17 + (-1.31)) \cdot 1 \qquad (9.4) \\
\hat{y}_1 &= \boxed{68.67} \quad \textit{Actual score}: 68.2
\end{aligned}$$

What we just saw for Learner_1 illustrates the concept of **shrinkage**, also referred to as "pooling factor". Hierarchical models will shrink individual values towards the mean. The amount of shrinkage will depend on the

```
predict(fit_lmer2,
        newdata = tibble(Feedback = "Explicit correction",
                         Task = "Task B",
                         ID = "Learner_1",
                         Item = "3"))
```

CODE BLOCK 52 Predicting Scores Using predict()

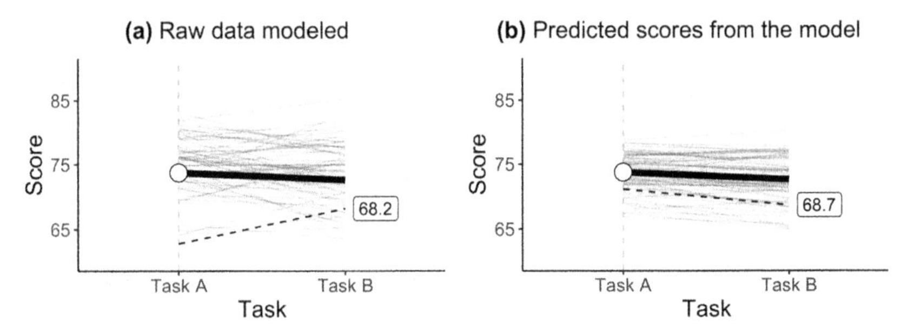

FIGURE 9.4 Shrinkage from Hierarchical Model

group-level variance in the data (among other factors). The intuition here is that by doing so, models are better able to generalize their predictions to novel data—see, for example, Efron and Morris (1977). After all, our sample of participants is random, and we shouldn't consider their individual values as accurate representations of the population as a whole—this is tied to our discussion on overfitting models in chapter 7.

To visualize what shrinkage does to our predicted values, let's inspect Fig. 9.4. First, Fig. 9.4a plots our actual data—the dashed black line represents the scores for Learner_1, assignment 3 (Item = 3), for both tasks.[5] Now look at Fig. 9.4b, which plots the estimated scores from fit_lmer2. Notice how all the gray lines, which represent individual participants, are closer to the grand mean (thick solid line). The dashed line representing Learner_1 is significantly affected here, as it is pulled towards the overall trend (thick line). Why is there so much shrinkage here? The answer is simple: because we're looking at a single assignment score from a single participant, our sample size here is minuscule (n = 1), and therefore its impact on the overall trend is very small. Simply put, such a minuscule sample size is not strong enough to avoid a lot of shrinkage.

Finally, let's compare fit_lmer1 and fit_lmer2 to see whether adding a random slope actually improves the model in question. We can do that by using the anova() function, which we discussed back in §6.3. If you run anova(fit_lmer1, fit_lmer2), you will see that fit_lmer2 is statistically better (e.g., its AIC is lower than that of fit_lmer1). Therefore, between fit_lmer1 and fit_lmer2, we would report fit_lmer2.

> **REPORTING RESULTS**
>
> To analyze the effect of feedback and task, we ran a hierarchical linear model with by-participant random intercepts and slopes (for task), and by-item random intercepts. Results show that both feedback ($\hat{\beta} = 2.91, SE = 0.99$) and task ($\hat{\beta} = -1.17, SE = 0.54$) have an effect on participants' scores—see Table 9.2.

Notice that we don't normally report the random effects themselves—although you may want to add them to your statistical table, so the table provides a more comprehensive output of the model. Remember: what we really care about are the fixed effects. Table 9.2 reports our fixed effects and R^2 values (from r.squaredGLMM()). At the top of the table, we also have the specification of the model.[6] The table may seem too minimalistic, but remember that estimates, SEs, and *t*-values are all we need in the end. With these values, we could compute 95% confidence intervals for all three fixed effects in fit_lmer2—you could also add them to the table.

Instead of a table, we could again plot the model's estimates just like we did before (either manually or using the plot_model() function from the sjPlot package). Fig. 9.5 is an intuitive way to visualize both fixed and random effects from fit_lmer2—intercept and slopes are plotted separately due to their different locations on the *x*-axis. The solid lines represent the 95% confidence interval of the estimates for our fixed effects, and the semitransparent circles represent by-participant effects (for our intercept and for the slope of Task)—compare the figure to Table 9.2, as they both show the same fixed effects. Because our model has no random effects for Feedback, no circles are shown for this variable in the figure—code blocks 53 and 54 show how to create Fig. 9.5 manually.

Finally, you should also inspect a histogram of the distribution of residuals (just like you would for a non-hierarchical model). Your residuals should follow a normal distribution (see discussion in chapter 6).

TABLE 9.2 Model Estimates and Associated SEs

	Score ~ Feedback + Task + (1 + Task | ID) + (1 | Item)		
	Estimate ($\hat{\beta}$)	*SE*	t-*value*
(Intercept)	73.74	3.58	20.63
Feedback (Recast)	2.91	0.99	2.93
Task (Task B)	−1.17	0.54	−2.14
	Conditional R^2 = 0.75; Marginal R^2 = 0.02		

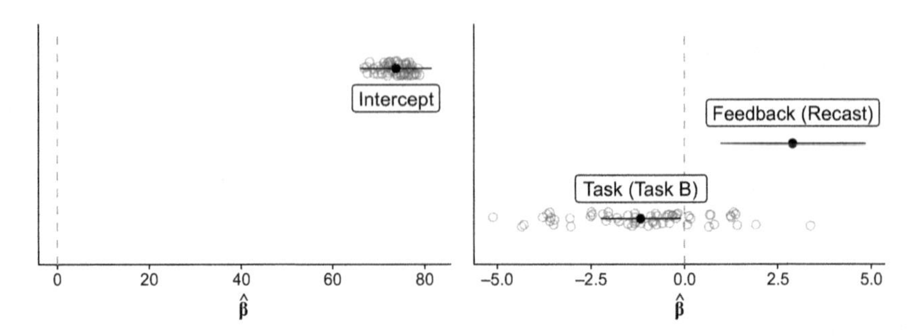

FIGURE 9.5 Plotting Estimates for Fixed and Random (by-Participant) Effects

As you can see, nothing substantial changes on the surface: we interpret and report our estimates the same way. This should make sense, since a hierarchical linear regression is a linear regression after all. Likewise, you can run hierarchical logistic regressions (glmer() function in the lme4 package) and hierarchical ordinal regressions (clmm() function in the ordinal package), as already mentioned. The take-home message is this: once you understand logistic regressions and hierarchical models, you will also understand a hierarchical version of a logistic regression—so everything we've discussed thus far is cumulative.

Given all the advantages of hierarchical models, it's safe to say that you should virtually always use them instead of non-hierarchical models—hopefully the discussion around Fig. 9.2 made that clear (see, e.g., discussion in McElreath 2020). Hierarchical models are more complex, but that's not typically a problem for computers these days. Your role as the researcher is to decide what you should include in your model (both as fixed effects and as random effects, recall the model comparison earlier using anova()). Ultimately, your decisions will be guided by your own research questions, by your theoretical interests, by the structure of your data, and by the computational complexity involved in the model itself—see, for example, Barr et al. (2013) and Bates et al. (2015a).

If your model is too simplistic (random intercepts only), it will run very quickly and you will likely not have any convergence issues or warning messages. However, because your model is too simple, Type I (or Type II) error is likely. On the other hand, if your model is too complex, it may take a while to run, and Type II error is a potential problem—that is, being too conservative makes you lose your statistical power. On top of that, complex hierarchical models can fail to converge—see, for example, Winter (2019, pp. 265–266). For that reason, if you start with a complex model, you may have to simplify your model until it converges—you will often see people reporting their "maximal converging model", which basically means they are using the most complex model that converged. Fortunately, the R community

R code

```
# Extract random effects:
rand = tibble(ID = row.names(coef(fit_lmer2)$ID),
              Intercept = coef(fit_lmer2)$ID[[1]],
              TaskB = coef(fit_lmer2)$ID[[3]])

rand = rand %>%
  pivot_longer(names_to = "Term",
               values_to = "Estimate",
               cols = Intercept:TaskB) %>%
  mutate(Term = factor(Term,
                       labels = c("Intercept", "Task (Task B)")))

# Extract fixed effects:
FEs = fixef(fit_lmer2)

# Extract lower and upper bounds for 95% CI:
CIs = confint(fit_lmer2)

# Combining everything:
fixed = tibble(Term = row.names(CIs[c(6,7,8),]),
               Estimate = FEs,
               Lower = c(CIs[c(6,7,8), c(1)][[1]],
                         CIs[c(6,7,8), c(1)][[2]],
                         CIs[c(6,7,8), c(1)][[3]]),
               Upper = c(CIs[c(6,7,8), c(2)][[1]],
                         CIs[c(6,7,8), c(2)][[2]],
                         CIs[c(6,7,8), c(2)][[3]]))

# Change labels of factor levels:
fixed = fixed %>%
  mutate(Term = factor(Term, labels = c("Intercept", "Feedback (Recast)",
                                        "Task (Task B)")))
```

CODE BLOCK 53 Plotting Fixed and Random Effects Manually (Part 1)

online is so huge that you often find effective help right away simply by pasting error messages on Google.

9.3 Additional Readings on Regression Models

Regression analysis, especially hierarchical models, is a huge topic, and it may take a while to have even some basic understanding of how these models work. Numerous books and articles exist which focus exclusively on regression analysis or on a particular type of regression. This chapter, much like chapters 6–8, has only scratched the surface, but it provides the basic set of tools for you to get started effectively. By using only what we have been discussing in this book, you will already be significantly ahead of the curve in SLA research—which is the main goal here (see Plonsky 2013 for a review of study quality in second language acquisition).

R code
```
library(cowplot) # to combine multiple plots (see below)
# Plot effects
# Plot intercept:
estimates_intercept = ggplot(data = fixed %>%
                              filter(Term == "Intercept"),
                            aes(x = Term, y = Estimate)) +
  geom_pointrange(aes(ymin = Lower, ymax = Upper), size = 0.3) +
  geom_jitter(data = rand %>% filter(Term == "Intercept"), shape = 21,
              width = 0.1, alpha = 0.1,
              size = 2) +
  labs(x = NULL, y = expression(hat(beta))) +
  coord_flip() +
  scale_x_discrete(limits = c("Task (Task B)", "Feedback (Recast)", "Intercept")) +
  geom_hline(yintercept = 0, linetype = "dashed", alpha = 0.3) +
  geom_label(aes(label = Term), position = position_nudge(x = -0.4)) +
  theme_classic() +
  theme(axis.text.y = element_blank(),
        axis.ticks.y = element_blank())

# Plot slopes:
estimates_slopes = ggplot(data = fixed %>%
                           filter(Term != "Intercept"),
                         aes(x = Term, y = Estimate)) +
  geom_pointrange(aes(ymin = Lower, ymax = Upper), size = 0.3) +
  geom_jitter(data = rand %>% filter(Term != "Intercept"), shape = 21,
              width = 0.1, alpha = 0.1,
              size = 2) +
  labs(x = NULL, y = expression(hat(beta))) +
  coord_flip() +
  scale_x_discrete(limits = c("Task (Task B)", "Feedback (Recast)", "Intercept")) +
  geom_hline(yintercept = 0, linetype = "dashed", alpha = 0.3) +
  geom_label(aes(label = Term), position = position_nudge(x = 0.4)) +
  theme_classic() +
  theme(axis.text.y = element_blank(),
        axis.ticks.y = element_blank())

# Combine both plots (this requires the cowplot package):
plot_grid(estimates_intercept, estimates_slopes)

# ggsave(file = "figures/estimates.jpg", width = 7, height = 2.5, dpi = 1000)
```

CODE BLOCK 54 Plotting Fixed and Random Effects Manually (Part 2)

If you wish to learn more about regression models, or if you are interested in more advanced topics related to them, you should definitely consult sources such as Gelman and Hill (2006), who offer comprehensive coverage of regression models—you should also consult Barr et al. (2013) on random effects in regression models. A more user-friendly option would be Sonderegger et al. (2018) or Winter (2019, ch. 14–15)—who also offers some great reading recommendations. Additionally, Cunnings (2012; and references therein) provides a brief overview of mixed-effects models in the context of second language research. Finally, see Matuschek et al. (2017) and references therein for a discussion on Type I and Type II errors in hierarchical models.

In the next chapter, we will still work with hierarchical regression models. As a result, our discussion will continue—this time within the framework of Bayesian statistics.

9.4 Summary

In this chapter we explored hierarchical models and their advantages. Thus far, we have covered three types of generalized linear models, namely, linear, logistic, and ordinal. All three models can be made hierarchical, so our discussion in this chapter about hierarchical linear models can be naturally applied to logistic and ordinal models in chapters 7 and 8, respectively. You now have all the basic tools to deal with continuous, binary, and ordered response variables, which cover the vast majority of data you will examine in second language research.

- Hierarchical models are more powerful than non-hierarchical models. Because most research done in second language research contains grouped factors (such as the hypothetical feedback study examined here), you should always favor hierarchical models in your analyses.
- As usual, running these models in R is straightforward—as is interpreting their fixed effects. Even though the models are underlyingly very complex, what we care about are estimates (effect sizes)—which we have been examining since chapter 6.
- Defining which random effects your model should have can be a difficult task, especially because not all random effects are relevant (or even possible): you have to fully understand the structure of your data as well as the design of the study. But this is a good thing: after all, you *should* know your own data really well.
- Most functions we had been using for non-hierarchical models are still useful for hierarchical models. For example, we still use `display()` or `summary()` to visualize outputs and `predict()` to predict new data given a model fit. Again: these models are just more complex version of the models we were using before.
- Parts of our discussion were focused on the random effects of our model (not only their interpretation but also their actual estimates). Most of the time you don't care so much about that part of the model, however. In other words, what we did earlier is more than what you will normally do. On the other hand, you will normally try more than two models, and you will most likely explore different variables in your models.

9.5 Exercises

Ex. 9.1 Hierarchical Models for Categorical Responses

1. In this exercise, we will run a hierarchical logistic regression. Using the `glmer()` function, rerun `fit_glm4` from chapter 7 with by-item and

by-speaker random intercepts—call the model fit_glmer1. Inspect the output using display() and compare it to the output of fit_glm4. You will need to source dataPrepCatModels.R, and you should also make sure that the reference levels for Condition and Proficiency are NoBreak and Nat, respectively. This will be a good opportunity to practice changing reference levels. Alternatively, you can rerun code blocks 36 and 38, which relevel both variables—and which should be located in logModels.R, that is, in the same directory we are currently working from.

a) How do the standard errors of the estimates in both models differ? Do our conclusions about our predictors change with fit_glmer1?
b) Explain why it would (or would not) make sense to add a random slope for Proficiency.

2. Using the clmm() function, rerun fit_clm1 from chapter 8 with by-item random intercepts and with by-speaker random slopes for Condition. Call the model fit_clmm1. Print the output for the two models using summary() (display() won't work with clm models).

   a) How do estimates and standard errors compare between the two models? Are the estimates for $\hat{\tau}$ different?
   b) Not all models that are complex will be necessarily better—especially if the complexity is not supported by the data (hence the importance of visualizing the patterns in our data prior to running models). Using the anova() function, compare fit_clm1 and fit_clmm1. Does fit_clmm1 offer a better fit (statistically speaking)? Which model has the lower AIC? Why do you think that is? On this topic, you may want to take a look at discussion in Matuschek et al. (2017) for future reference.

3. Create a figure for the rc dataset that mirrors Fig. 4.3 in chapter 4—except for the bars, which will be replaced by point ranges here. The figure will only include Spanish speakers, so make sure you filter the data in the first layer of ggplot()—as a result, only two facets will be present (cf. Fig. 4.3). On the *y*-axis, add as.numeric(Certainty), since we will treat certainty as a continuous variable here. On the *x*-axis, add Condition. Then use stat_summary() to plot mean certainty and standard errors. You should add by-speaker lines representing each participant's mean certainty level. How does the figure help you understand the previous question?

Notes

1. Note that we spell lme4 as L-M-E-4 (the first character is not a number).
2. You can use the package in question to extract R^2 values from logistic models as well—hierarchical or not. However, recall that linear and logistic models operate differently

(chapter 7); see Nakagawa and Schielzeth (2013). These are therefore "pseudo-R^2" values for logistic models.

3. Simply install and load the package, and then rerun your model. After that, print the output using summary() and you will see *p*-values.
4. You can run the same model without the correlation parameter by using double pipes ("||") in the model specification: (1 + Task || ID)—but first you will need to create a numeric version of Task. See Sonderegger et al. (2018, §7.8.2) for a discussion on adding a correlation to hierarchical models.
5. Notice that every gray line represents a single participant's score, aggregated from all five assignments in each task. The dashed line for Learner_1, however, *only* considers the scores for Item = 3.
6. You may or may not keep it there for your own models when you report them. Your text should already be explicit about the structure of the model you are employing, but I find that having the structure in the table as well can be useful for the reader.

10
GOING BAYESIAN

Thus far, all the models we have run in this book are based on Frequentist data analysis. You may not be familiar with the term *Frequentist*, but every time you see *p*-values you are looking at Frequentist statistics—so you are actually familiar with the concept. But while *p*-values seem to be everywhere in our field, not all statistical analyses must be Frequentist. In this chapter, we will explore *Bayesian* statistics, a different approach to data analysis. There are numerous differences between Frequentist and Bayesian inference, but you'll be happy to know that all the models we have run so far can also be run using Bayesian statistics.

The first important difference between Frequentist and Bayesian statistics can be contextualized in terms of probabilities and how we conceptualize them. For example, what's the probability that a given learner will choose low attachment in our hypothetical relative clause study? A Frequentist would define the probability based on several samples. He or she would collect data and more data, and the long-term frequency of low attachment would reveal the probability of interest. As a result, if after ten studies the average probability of low attachment is 35%, then that's the probability of a learner choosing low attachment. For an orthodox Frequentist, only repeatable events have probabilities.

A Bayesian would have a different answer to the question. He or she would first posit an initial assumption about the true probability of choosing low attachment. For example, knowing that Spanish is biased towards high attachment, the probability of choosing low attachment is likely below 50%. Then data would be collected. The conclusion of our Bayesian statistician would be a combination of his or her prior assumption(s) and what the data (i.e., the evidence) shows. We use Bayes's theorem[1] to evaluate this combination and to reach a conclusion about the estimated probability of choosing low attachment.

Here's the thing: we use Bayesian reasoning every day. When you are about to leave the house, you have to consider whether it might rain so you can dress accordingly and take an umbrella with you. To do that, you check your weather app, sure, but you also take into account that the weather in your region is quite unstable, so you can never trust weather forecasts 100%—this is your prior knowledge. Besides consulting your app, you probably also look outside. You are basically gathering data, considering your prior experiences, and then deciding whether to take an umbrella with you. If you now decide to move to another region in the country, you will lack important prior experience about the weather there—you may ask your new neighbors, who are more experienced and therefore have more informed priors about the local weather. We use our prior experiences to assess future outcomes on a regular basis: when we visit new places, meet new people, when we make important decisions, and so on.

WHY SHOULD WE CARE ABOUT PRIORS IN RESEARCH? For one, it is obvious that we should *not* ignore all the body of scientific knowledge that exists out there. As Sarton (1957) puts it, "the acquisition and systematization of positive knowledge are the only human activities which are truly cumulative and progressive". In second language research, for example, we know from numerous previous studies that learners transfer patterns from their first language onto a second language—for example, White (1989) and Schwartz and Sprouse (1996). As a result, virtually no one assumes that second language learners start from scratch, with no grammar at all.

When we run a typical (i.e., Frequentist) statistical model, we cannot inform the model that certain patterns are more expected than others. Simply put, our models ignore everything we know about linguistics and language acquisition: they are by definition naïve and completely disregard language transfer in this case. This means that our Frequentist models cannot accommodate our knowledge that (i) Spanish and English seem to have different attachment preferences and (ii) learners transfer patterns from their native language. Clearly, this is not ideal.

Suppose that 100 previous studies point to a given negative effect. Then a new study is run, and its results contradict all previous studies, showing a statistically significant *positive* result ($p < 0.05$). Should we discard 100 negative effects in favor of a single positive effect? We should not. Clearly, a single study cannot possibly have such a huge effect on our conclusions. If we can inform our statistical model that 100 previous studies agree on a negative effect, then it will take *a lot* of data for our results to contradict all previous studies. This should all be intuitive: to replace a solid body of scientific knowledge, we need very convincing evidence (not just a single study where $p <$

0.05). To quote Carl Sagan, "extraordinary claims require extraordinary evidence".[2]

Given that our prior knowledge matters and that Bayesian inference is much older (18th century) than Frequentist inference (20th century),[3] why haven't we been using Bayesian methods all this time? One reason is computational power, which was a serious limitation until not long ago. These days, however, even a cheap laptop has considerable computational power compared to the computers of the past.

In this chapter, you will learn how to run the models we've run so far using Bayesian estimation. We will also see how to interpret the models, report and plot our results, and specify mildly informed priors that incorporate our linguistic knowledge into our statistical models. By the end of the chapter, you will be able carry out your own analysis using very familiar coding from previous chapters.

The overall structure of this chapter is very similar to that of previous chapters. We first examine the basics of Bayesian statistics (§10.1) and then run different models in R (§10.4). This time, however, I will introduce you to a new type of file (.RData) in §10.2, before we start running our models. We will also need to get your RStudio ready to run Bayesian models (§10.3).

There are numerous excellent books out there that focus solely on Bayesian statistics, and this chapter could not possibly do justice to such a huge topic. In addition, we won't get into the mathematical details of Bayesian models here, as our main goal is to apply such models to our data. For those reasons, if you wish to explore Bayesian statistics in more detail, refer to the reading suggestions at the end of this chapter.

10.1 Introduction to Bayesian Data Analysis

As we start our discussion on Bayesian inference, it may be helpful to briefly review a key concept in traditional (Frequentist) statistics, namely, *p*-values (§1.3.1). Later in this section we will also discuss confidence intervals.

As you know, *p*-values give us the probability of the data given a particular statistic. In other words, our models will estimate an effect $\hat{\beta}$ and will consider all possible datasets that could be generated—keeping $\hat{\beta}$ fixed. In that context, our *p*-value will tell us the probability of observing data that are at least as extreme as the data we have—given $\hat{\beta}$. We can simplify that by saying that a Frequentist model gives us the probability of the data given a parameter value π (under the null hypothesis), or $P(data \mid \pi)$ assuming H_0. Notice that *p*-values are *not* the probability of the parameter we are interested in. As a result, when we run a Frequentist linear model, we don't know the probability of our $\hat{\beta}$ estimate.

In contrast with Frequentist statistics, Bayesian statistics gives us the probability of the parameter given the data, or $P(\pi \mid data)$—which we call *posterior*. Here, it's the dataset that is fixed, and different parameter values are considered—notice that this is conceptually very different from what we get in Frequentist statistics. Ultimately, we are much more interested in parameters than in data.

To calculate our posterior, $P(\pi \mid data)$, we use Bayes's theorem, shown in 10.1. The probability of a parameter value given the data (posterior) equals the probability of the data given that parameter value (likelihood), times the probability of the parameter value (prior), divided by the total probability of the data.

$$\overbrace{P(\pi|data)}^{posterior} = \frac{\overbrace{P(data|\pi)}^{likelihood}\ \overbrace{P(\pi)}^{prior}}{P(data)} \tag{10.1}$$

The intuition behind Bayes's theorem is simple: the stronger our priors are, the more data we need to see to change our posterior. Consider this: the more we believe in something, the more evidence we require to change our minds (if we blindly believe in something, no evidence can convince us that we're wrong). Here, however, our priors should be informed not by what we wish to see, of course, but rather by the literature—see Gelman (2008a) for common objections to priors in Bayesian inference.

Let's see 10.1 in action—we will use our hypothetical relative clause study to guide us here. First, we establish our prior probability that learners will choose low attachment, let's call that probability $P(\pi)$—π is our parameter of interest, which goes from 0 to 1, as shown in Fig. 10.1. Because these learners are speakers of Spanish, and because we know the literature, we assume that they are

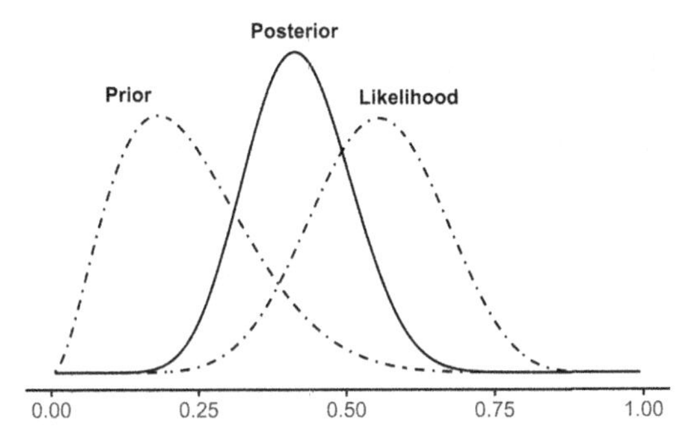

FIGURE 10.1 Bayes's Rule in Action: Prior, Likelihood, and Posterior Distributions

more likely to choose high than low attachment, so we expect $\pi < 0.5$, or, more explicitly, $P(Low = 1) < 0.5$. Notice that everything in Fig. 10.1 is a distribution (more specifically a beta distribution)[4]: we don't see single-point estimates in the figure.

Given what we know about the topic, let's assume that our prior probability will peak at ≈ 0.18. This is like saying "we believe *a priori* that the most probable value for π is approximately 0.18, that is, Spanish speakers will choose low attachment 18% of the time. However, π follows a distribution, so other values are also probable—they're simply less probable relative to the peak of the distribution". We then run our experiment and collect some data. What we observe, however, is that participants chose low attachment over 50% of the time: this is our likelihood line in Fig. 10.1. To be more specific, we collected 13 data points, 5 Low and 8 High. What should we conclude about π?

As you can see, our posterior in Fig. 10.1 is a compromise between our prior and the data observed. We started out assuming that the most probable value for π was 0.18. Our data, however, pointed to $\pi = 0.55$ as the most probable value for π. Our posterior distribution then peaks at $\pi = 0.41$. Again: we are talking about probability distributions, so when we say "the most probable value for parameter π", we are essentially talking about the peak of the distribution.

Notice that our prior distribution in Fig. 10.1 is relatively wide. The width of our distribution has an intuitive meaning: it represents our level of certainty. Sure, we believe that Spanish learners will choose low attachment only 18% of the time. But how sure are we? Well, not too sure, since our prior distribution clearly extends from zero to over 50%. Intuitively, this is equivalent to saying "we have some prior knowledge based on previous studies, but we are not absolutely sure that 0.18 is the true value of π".

Let's see what happens if we are much more certain about our prior beliefs regarding π. In Fig. 10.2, our prior distribution is much narrower than that of Fig. 10.1—the peak of our prior distribution is now ≈ 0.23. Because we are now more certain about our prior, notice how our posterior distribution is much closer to the prior than it is to the likelihood distribution. Here, we are so sure about our prior that we would need a lot of data for our posterior to move away from our prior.

In Figs. 10.1 and 10.2, we assumed our data came from only 13 responses (5 Low and 8 High). What happens if we collect ten times more data? Let's keep the same ratio of Low to High responses: 50 Low and 80 High. In Fig. 10.3, you can see that our posterior is now much closer to the data than it is to our prior (which here is identical to the prior in Fig. 10.1). The intuition here is simple: given enough evidence, we have to reallocate our credibility. In other words, after collecting data, we discovered that we were wrong.

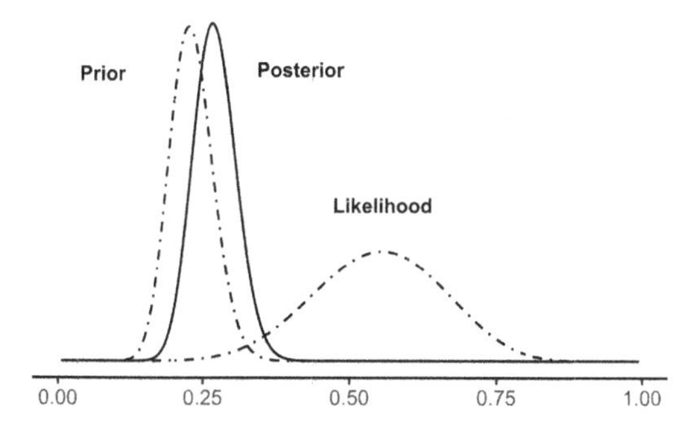

FIGURE 10.2 How a Strong Prior Affects our Posterior

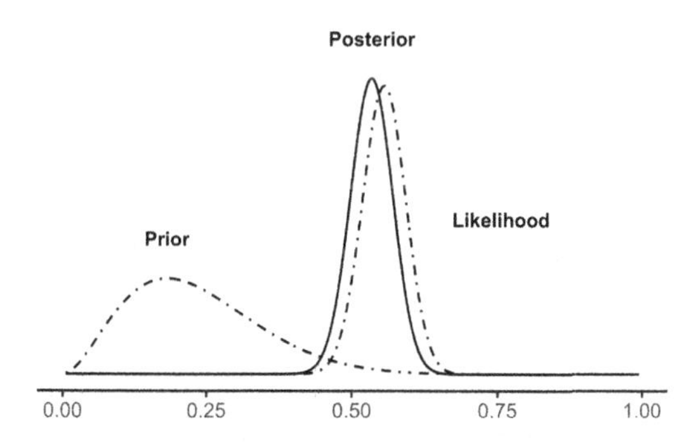

FIGURE 10.3 How a Larger Sample Size Affects our Posterior

We saw earlier that Bayesian inference gives us $P(\pi | data)$, while Frequentist inference gives us $P(data | \pi)$. Our discussion also revealed another crucial difference between these two methods: everything in Figs. 10.1–10.3 is a distribution of credible parameter values. When we run linear models such as those in chapters 6, 7, and 8 using Bayesian estimation, our estimates will be posterior distributions. Here's what this means in practice: before, in our Frequentist models, we reported a point-estimate, our $\hat{\beta}$. In a Bayesian model, our $\hat{\beta}$ will be an actual distribution of values, not a single point-estimate. You may be wondering how we can report distributions, but the answer is actually very simple: we can report the value that is at the top of the distribution (its mode) or the mean of the distribution (depending on how symmetrical the distribution is). That value is the most probable parameter value given the data, so that's our "best candidate" for $\hat{\beta}$. In addition to that, we can provide some interval (e.g., 95% interval) that conveys how wide the posterior distribution actually is.

It should be clear by now that we don't have *p*-values in Bayesian models, since we're not calculating the probability of the data anyway. On top of that, our result is an entire (posterior) distribution of credible parameter values given the data—that's much more informative, comprehensive, and meaningful. But you may be wondering whether the interval I mentioned earlier is equivalent to confidence intervals (§1.3.3). The short answer is: no.

We refer to intervals in posterior distributions as *credible intervals*, not confidence intervals (to avoid confusion, since they do mean different things). Don't worry: understanding them is much easier than understanding confidence intervals (do you actually remember what confidence intervals are...?). It's not surprising that many people misinterpret Frequentist confidence intervals (see Kruschke and Liddell 2018 and references therein regarding misinterpretations involving Frequentist concepts). Let's briefly recap what they mean before we approach Bayesian credible intervals.

Frequentist 95% confidence intervals (CIs) are based on hypothetical future sampling. If we repeated an experiment *n* times, then the true parameter value would be within our confidence interval 95% of the time. Clearly, this is a very abstract concept: first, we will likely not repeat our experiment multiple times. Second, the 95% confidence interval will be *different* every time we replicate the experiment.

Now let's turn to credible intervals in Bayesian statistics. As an example, let's use the posterior distribution of Fig. 10.1. We can calculate different types of intervals here, but we will focus on a specific interval commonly referred to as the *highest density interval*, or HDI for short.[5] This interval has a very intuitive interpretation: it contains the most probable parameter values for π. Fig. 10.4 plots the posterior distribution as well as its 95% HDI (shaded area).[6]

The horizontal point range at the bottom of Fig. 10.4 concisely conveys all we need to know: the mean of the posterior as well as its 95% HDI: $\pi = 0.42$, 95% HDI = [0.25, 0.59]. In other words, once we take into account our prior and the hypothetical data discussed earlier, we conclude the most credible value for π is 0.42, that is, Spanish learners have a 42% probability of choosing low attachment. However, within our 95% HDI we also have 0.5 as a credible statistical value for π,[7] so we cannot conclude with absolute certainty that a bias exists: we can only argue that the evidence points to a bias, but more data is needed. Notice that there *is* a bias, as we can see in the posterior distribution, which is clustered *below* 0.5 (even though 0.5 is included in the HDI, it's not as credible as the peak of the distribution). In a Frequentist analysis, the confidence interval would include 0.5 (0 log-odds), and we would simply conclude "there's no effect, since $p > 0.05$". Our conclusion here must be more nuanced.

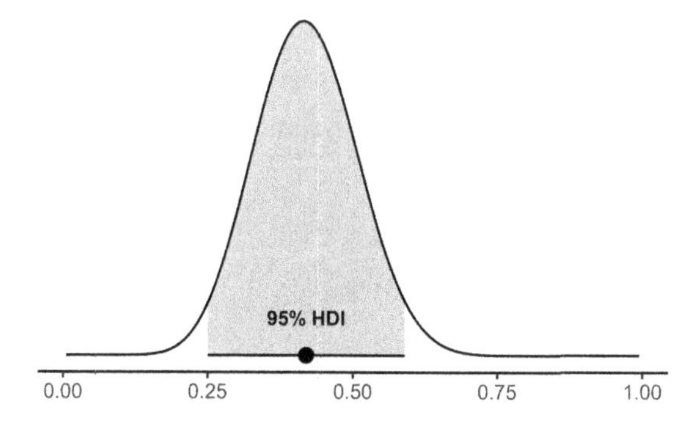

FIGURE 10.4 95% Highest Density Interval of Posterior from Fig. 10.1

Unlike Frequentist confidence intervals, the HDI has a straightforward interpretation: values inside the interval are more credible than values outside the interval. In addition, notice that HDIs are actual probability distributions. Consequently, values closer to the mean are *more* statistically credible given the data than values that are far from the mean in the interval (cf. confidence intervals, which are *not* probability distributions).

When we report the mean posterior distribution for π earlier, the estimate reported is merely the peak of an entire distribution of credible values. Considering distributions instead of single values is not only more comprehensive and informative, it's also more realistic: clearly, 0.418 is also a highly credible parameter value for π given the data—hence the importance of also reporting HDIs. We could now adjust our beliefs and update our prior for π with our posterior distribution, so the next time we run the same experiment our expectations will be different given the evidence we have seen so far.

10.1.1 Sampling From the Posterior

The example discussed in Figs. 10.1–10.3 is very simple: we only consider one parameter, π. In more realistic models, like the models we use in chapter 9, we will have multiple parameters (e.g., multiple predictor variables). The problem is that when we consider a model with n parameters, the parameter space is n-dimensional. Assume, for example, that we have seven parameters—this is a relatively low number, especially considering hierarchical models. Next, assume that we wish to consider 1,000 parameter values for each parameter (we could assume more, of course). The result is a joint distribution that has $1{,}000^7$ combinations of parameter values—our computers can't handle that.

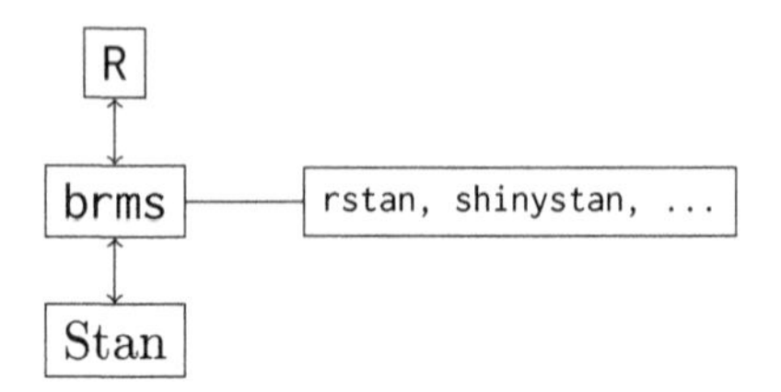

FIGURE 10.5 Bayesian Models in R Using Stan

The solution to the problem is to sample values from the posterior distribution without actually solving the math—this is essentially what we will do in R. Sampling is at the core of Bayesian parameter estimation, which means we need an efficient sampling technique. One powerful sampling method is called Monte Carlos Markov chain (**MCMC**). In this book, we will rely on a specific MCMC method, called *Hamiltonian Monte Carlo*, or HMC. The details of these techniques are fascinating (and complex), but we don't need to get into them to be able to run and interpret Bayesian models—see Kruschke (2015).

The language (or platform) we will use to run our models is called Stan (Carpenter et al. 2017), named after Polish scientist Stanislaw Ulam.[8] We won't actually use Stan directly (but you will be able to see what its syntax looks like if you're curious). Instead, we will use an R package called brms (Bürkner 2018), which will "translate" our instructions into Stan. Fig. 10.5 illustrates the process: we will soon install and load brms, which itself loads some dependent packages, such as rstan (Stan Development Team 2020)—wait until §10.3 before installing brms.

The main advantage of brms is that it lets us run our models using the familiar syntax from previous chapters. The model is then sent to Stan, which in turn compiles and runs it. Finally, the output is printed in R. On the surface, this will look very similar to the Frequentist models we ran in previous chapters—in fact, if I hadn't told you about it you would probably not even notice at first sight that there's something different happening. But below the surface *a lot* is different.

You may be wondering how we can know whether the sampling (or our model as a whole) worked. In other words, how can we inspect our model to check whether it has found the appropriate posterior distribution for our parameters of interest? When we sample from the posterior, we use multiple *chains*. Our chains are like detectives looking for the parameter values (π) that are most credible given the data. The detectives begin a random walk in the parameter space, and as their walk progresses, they cluster more and more around parameter values that are more likely given the data that we are modeling. By having multiple chains, we can easily compare them at the end of the process to see whether they have all arrived at (roughly) the same values. This will be easy to interpret when we inspect our models later in this chapter, when we will discuss model diagnostics in more detail.

Finally, let's briefly examine what the models we ran in previous chapters look like in a Bayesian framework. This will help you understand what is going on once we run our models later in the chapter. Let's pick the simplest model we have run so far. In 10.2, you can see a linear regression with an intercept and a single predictor, $\hat{\beta}_1$. This is the fundamental model/template we have been using in this book (generalized linear models).

$$y_i = \beta_0 + \beta_1 x_{i1} + \epsilon_i \tag{10.2}$$

In a Bayesian framework, we still use essentially the same model. In R, we will still specify our model as `Score ~ Hours`—but it's important to understand what's going on as we do that. In our Bayesian models, we specify the distributions assumed for each term that will be estimated—shown in 10.3. Let's go over it step by step. First, we assume that our response variable y_i follows ($\sim$) a normal (i.e., Gaussian) distribution, represented as $\mathcal{N}$. This is standard in linear regressions, since our response variable is continuous and assumed to be Gaussian.[9]

$$\begin{aligned}
y_i &\sim \mathcal{N}(\mu_i, \sigma) \\
\mu_i &= \boxed{\beta_0 + \beta_1 x_{i1}} \\
\beta_0 &\sim \mathcal{N}(0, 100) \\
\beta_1 &\sim \mathcal{N}(0, 100) \\
\sigma &\sim \mathcal{U}(0.001, 1000)
\end{aligned} \tag{10.3}$$

Once we define that our response variable follows a normal distribution, we have to define the mean and standard deviation of said distribution. What's the mean of the distribution? It's our linear model, shown in the second line of 10.3. Think about it for a second: we are saying that the score of a participant is continuous and comes from a normal distribution with mean μ. That mean, in turn, is defined as a linear model, so we can see how different explanatory variables (predictors) can affect the mean of the distribution from which scores are derived.

Now that we have a linear model, we need to know what distributions are assumed for the terms in the model. Here, our model has two terms, namely, an intercept (β_0) and a slope (β_1). We assume that the estimates from both terms also come from a normal distribution, only this time we are specifying the means of said distributions ourselves. Important: we are not saying that the *variable* (i.e., x_i) is normally distributed (after all, this could be a categorical/binary variable). What we are saying is that *the effect* (i.e., β) of the variable is normally distributed. Thus, β need not refer to a continuous variable *per se*. We are also specifying the standard deviation of the distributions.

Let's stop for a second to contextualize our discussion thus far. In fit_lm1, back in chapter 6, we ran Score ~ Hours, so it was a model with the same structure as the model exemplified in the specification given in 10.3. Back in chapter 6, the estimates for fit_lm1 were Intercept = 65 and Hours = 0.92. In the model specified in 10.3, we are assuming the intercept has a prior distribution centered around 0. If we were fitting fit_lm1, we would be off by a lot, since $\hat{\beta}_0 = 65$ in that model. However, notice that the standard deviation assumed here is 100, which is considerably wide. Therefore, the model would be relatively free to adjust the posterior distribution given the evidence (data) collected.

Notice that here we are being explicit about the standard deviations for β_0 and β_1 (i.e., 100). But we could let the model estimate them instead by specifying different priors for these standard deviations. This is an important point because in Frequentist models we are obligated to assume that variance will be constant (homoscedasticity). In Bayesian models, on the other hand, that doesn't have to be the case.

Finally, the last line in 10.3 tells the model that we expect σ to follow a noncommittal uniform distribution ($\mathcal{U}$). The simple model in 10.3 is estimating three parameters, namely, β_0, β_1, and σ. Of these, we're mostly interested in β_0 and β_1. Once we run our model on some data, we will be able to display the posterior distribution of the parameters being estimated. Fig. 10.6 illustrates what the joint posterior distribution of β_0 and β_1 looks like—both posteriors follow a normal distribution here. The taller the density in the figure, the more credible the parameter value. Thus, the peak of the two distributions represents the most likely values for β_0 and β_1 given the data, i.e., $P(\hat{\beta}_0|data)$ and $P(\hat{\beta}_1|data)$. We will see this in action soon.[10]

One of the main advantages of specifying our own model is that we can choose from a wide range of distributions to set our priors. Thus, we no longer depend on assumptions imposed by the method(s) we choose. At the

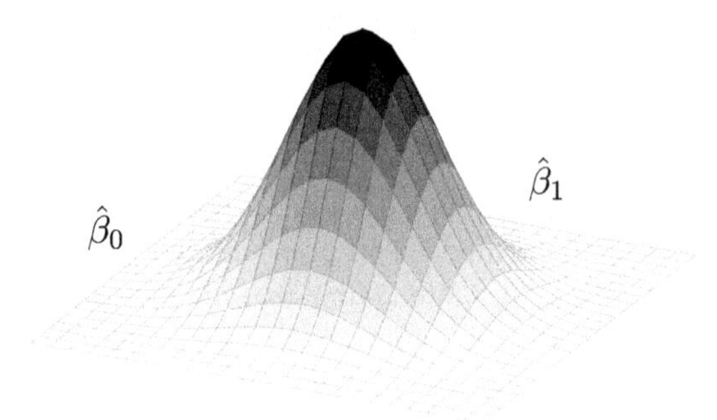

FIGURE 10.6 Illustrative Posterior Distributions for $\hat{\beta}_0$ and $\hat{\beta}_1$

same time, as we will see later, we can also let our model decide which priors to use. This is especially useful for people who are new to Bayesian models and who are not used to manually specifying models. Ultimately, however, if you wish to maximize the advantages of a Bayesian model, you should consider customizing priors based on previous research.

To illustrate my point, allow me to use photography as an analogy. Professional cameras give you a lot of power, but they require some knowledge on the part of the user. Fortunately, such cameras have an "auto" mode, which allows you to basically point and shoot. Hopefully, once you start shooting with a professional camera, you will slowly understand its intricacies, which in turn will lead you to better explore all the power that the camera has to offer. However, even if you decide to use the auto mode forever, you will still take excellent photos. What we are doing here is very similar: Bayesian models offer *a lot*, but you don't have to be an expert to use them—the "auto mode" we will see later gives you more than you expect (and more than enough to considerably advance your data analytical techniques). Naturally, if you decide to learn more about Bayesian statistics, you will get better at customizing your model to your needs.

10.2 The RData Format

Running complex models can take time. Imagine the following situation: you run a hierarchical model (Frequentist or Bayesian) with several predictors; the process takes a couple of minutes, and at the end you have a new object in your workspace (i.e., a model fit). But objects are not files: once you close RStudio, that object is gone. What happens if in the future (e.g., five months from now) you want to check the model again? You need to open RStudio and rerun the model, which will again take a couple of minutes. This is clearly not ideal: we want to be able to save our model in a file, so we don't need to rerun it every time we close and open RStudio (or clean our workspace using `rm(list = ls())`).

Let me introduce you to `RData` files, the best way to save objects in R—*any* object. A single `RData` file can hold your model fits, your data frames or tibbles, or any other object you create in your script. As a result, if you load that single file at a later date, your entire workspace will be populated with all the objects you saved in it—including complex model fits. Once you know `RData` files exist, there's no reason not to use them.

Speed is not the only reason you should use `RData` files. They also improve your organization, since you don't have to rerun multiple scripts to recreate different objects, for example. On top of that, `RData` also compresses your data. To give you a simple example: the Portuguese Stress Lexicon (Garcia 2014), which I developed for Garcia (2017), has 154,610 rows and 62 columns. The `csv` version of the lexicon is nearly 55 MB. The `RData` version of the same lexicon is 9.4 MB.

Another advantage of RData files is their ability to also save metadata. For example, we often change the class of columns in our datasets. You may have a character column that you transformed into an ordered factor to run an ordinal model in your study (we did exactly that in dataPrepCatModels.R). However, csv files have no "memory", and the next time you load your csv file, you will have to perform the same transformation—of course, you will simply rerun the lines of code that do that for you, or source a script where you prepare your data, which is what we have been doing. If you save your tibble in an RData file, however, it will "remember" that your column is an ordered factor, and you won't have to transform it again.

At this point you should be convinced that RData files are extremely useful. So how do we use them? To save objects in RData file, we use the save() function (e.g., save(object1, object2, ..., file = "myFiles.RData")). The file will be saved in your current working directory unless you specify some other path. To load the file later on, we use the load() function (e.g., load("myFiles.RData"). That's it. We will see RData in use soon, when we save our Bayesian model fits.

10.3 Getting Ready

As mentioned earlier, for this chapter you will need to install the brms package (Bürkner 2018). Once you run install.packages("brms"), RStudio will also install additional packages (Fig. 10.5). Unlike in previous chapters, brms will require some additional steps when it's installed for the first time. After installing brms, follow the steps that follow.

If you use *Linux:* Depending on your distribution, you will need to check the instructions or warnings/errors generated after running install.packages("brms"). The code here has been tested using Ubuntu 20.04.

If you use *Mac OS:* You will need to install Xcode from the App Store. The installation will likely take a while. After you have successfully installed both brms and Xcode, restart your computer. You should now be able to run all the models that we will explore in §10.4.

If you use *Windows:* You will need to install Rtools. Both the file and the necessary instructions can be found at https://cran.r-project.org/bin/windows/Rtools/. You may also need to install the ps package once you have installed both brms and Rtools.

Finally, let's stop for a second to create a new folder and some new scripts. Inside the models folder, create a folder called Bayesian. There, create an R Project called Bayesian.RProj and two scripts: plotsBayesianModels.R and bayesianModels.R. You will also need to copy our data files feedbackData.csv and rClauseData.csv and paste them inside our Bayesian folder, since we will use those data files. Likewise, create a copy of dataPrepLinearModels.R and dataPrepCatModels.R, and create a folder for figures called figures, as usual. Your Bayesian folder should have one R Project file, two csv files, two new scripts, and two old scripts that read and prepare our data files (in addition to the figures folder)—refer to Appendix D. As you run the code blocks that follow, more files will be added to your directory. Recall that you don't have to follow this file organization, but it will be easier to reproduce all the steps here if you do (plus, you will already get used to an organized work environment within R using R Projects).

10.4 Bayesian Models: Linear and Logistic Examples

In this section, we will run one Bayesian model for our study on feedback and one for our study on relative clauses, where we will see how to use specific priors—both models will be hierarchical (chapter 9). Because we have already plotted our data in previous chapters, here we will go straight to the models—but you should plot your data first and then work on your models. We will run, interpret, display, and report our models' estimates. We will also discuss some model diagnostics, which in turn will help us assess whether our models are doing what they should be doing. In the end, you will notice that our interpretation is not too different from what we're used to. Indeed, relative to the Frequentist models in previous chapters, it will be more intuitive to interpret the results, given our earlier discussion. Finally, there are multiple ways to visualize Bayesian estimates, as we are now dealing with *distributions* instead of single values for our $\hat{\beta}$s. We will explore some options.

10.4.1 Bayesian Model A: Feedback

We will start by running the Bayesian equivalent to fit_lmer2, which we ran in chapter 9 (see code block 50). In that model, we predicted participants' scores as a function of both Feedback and Task. Our model was hierarchical and had a random intercept by participant and by item, as well as a random slope for Task effects by participant.

Code block 55 shows how to run our model (lines 4–9), fit_brm1, and prints part of its output (13–21)—note that the output is generated by simply running fit_brm1. You should place this code block inside bayesianModels.R. Line 23 saves the model as an RData file, and line 24 shows how to load the model later, so you don't need to rerun the model. On my laptop, running[11] fit_brm1 took approximately 130 seconds—in comparison, running fit_lmer2 took 0.10 seconds. Saving the model as an RData file is therefore highly recommended!

Most of what you see in lines 4–9 will look familiar from previous chapters. The package we load in line 1, brms, is doing all the work for us by translating our familiar syntax into Stan. What's different here are lines 7–9. Line 7 is specifying that we're running a linear regression, which has a normally distributed response variable.[12] Line 8 is saving the model code (not the model *per se*,

R code

```
library(brms)

source("dataPrepLinearModels.R")
fit_brm1 = brm(Score ~ Feedback + Task +
                 (1 + Task | ID) + (1 | Item),
               data = longFeedback,
               family = gaussian(),
               save_model = "fit_brm1.stan", # this will add a file to your directory
               seed = 6)

fit_brm1

# Population-Level Effects:
#                Estimate Est.Error l-95% CI u-95% CI Rhat Bulk_ESS Tail_ESS
# Intercept         73.78      3.96    65.78    81.69 1.00     1349     1309
# FeedbackRecast     2.91      1.03     0.92     4.96 1.00     1784     2553
# TaskTaskB         -1.18      0.53    -2.25    -0.12 1.00     4226     3272
#
# Family Specific Parameters:
#       Estimate Est.Error l-95% CI u-95% CI Rhat Bulk_ESS Tail_ESS
# sigma     5.18      0.17     4.87     5.51 1.00     2547     2929

# save(fit_brm1, file = "fit_brm1.RData") # this will add a file to your directory
# load("fit_brm1.RData")
```

CODE BLOCK 55 Fitting a Bayesian Hierarchical Linear Regression

which is saved in line 23). What does that mean? Since brms is actually translating our instructions into Stan, we have the option to actually save those instructions (written for Stan) as a separate text file (.stan extension), which will be added to your working directory (Bayesian). You won't need this file here, but it is actually important for two reasons: first, you can later open it using RStudio and inspect the actual code in Stan (which will allow you to see how brms is specifying all the priors in the model, i.e., what the "auto model" looks like). Once you do that, you will be very grateful that we can use brms to compile the code for us—learning how to specify models in Stan can be a daunting task. A second advantage of saving the model specification is that we can later have direct access to the entire model (this can be a nice learning tool as well). We won't get into that in this book, but see the reading suggestions at the end of the chapter.

Line 9 helps us reproduce the results. Each time you run a model, a random number is used to start the random walk of the chains, so results will be slightly different every time (remember that we're sampling from the posterior). By specifying a seed number (any number you want), you can later rerun the model and reproduce the results more faithfully.

There are several additional/optional arguments missing from our brm() function in code block 55. For example, we can specify how many chains we want (the default is four). We can also specify whether we want to use multiple cores in our computer to speed up the process. Importantly, we can also specify the priors of our model. We will do that later. For now, we're letting brms decide for us (think of this as the "auto mode" mentioned earlier). If you want to check other arguments for the brm() function, refer to the documentation of brms—or simply run ?brm in RStudio. You can also hit Tab inside the brm() function to see a list of possible arguments in RStudio, as already mentioned.

Let's now interpret the actual output of the model, which will look slightly different compared to your output once you run the model yourself. Code block 55 only shows our main effects (population-level effects), but random effects will also be printed once you run line 11. First, we have our estimates and their respective standard errors. This is quite familiar at this point. Then we have the lower and upper bounds of our 95% credible intervals (as discussed in Fig. 10.1). For example, the most credible value for our intercept[13] is $\hat{\beta} = 73.78$, but the 95% credible interval ranges from $\hat{\beta} = 65.78$ to $\hat{\beta} = 81.69$—you should be able to picture a posterior distribution with these values, but we will visualize these estimates soon. Notice that we're given the probability of the parameter given the data, $P(\hat{\beta}_0|data)$, not the probability of the data given the parameter (i.e., p-value in Frequentist inference; §10.1).

The last three columns in the output of fit_brm1 help us identify whether the model is working properly (i.e., model diagnostics). First, Rhat ($\hat{R}$), which is also known as the Gelman-Rubin convergence diagnostic. $\hat{R}$ is a metric used to monitor the convergence of the chains of a given model. If $\hat{R} \approx 1$, all our chains are in equilibrium—see Brooks et al. (2011). All our $\hat{R}$ in fit_brm1 are exactly 1, so we're good here—an $\hat{R} = 1.2$ is a bad sign.

Finally, we have two ESS columns (*effective sample size*, typically represented as $\hat{n}_{eff}$). This has nothing to do with the sample size in our study; it is instead the number of steps that are uncorrelated in our sampling. If our chains get stuck in a single location before moving on, the steps taken in that location are all correlated and therefore are less representative of the walk as a whole. Simply put, the higher the number, the better—higher numbers indicate that more uncorrelated steps were taken in the posterior estimation for a given parameter.

How can we visualize our Bayesian model? Throughout this book, I have been emphasizing the importance of data visualization to better understand our data. For Bayesian models, data visualization is even more crucial. Plotting is the best way to check whether our model has converged and to examine the posterior distributions of our estimates. Let's start with a *trace plot*, which we can use to inspect our chains. This is not the type of plot you would add to your paper, but it's useful to check it to make sure that all the chains arrived at the same location in the parameter space.[14] If all four chains in fit_brm1 converged towards the same parameter values, that's a good sign.

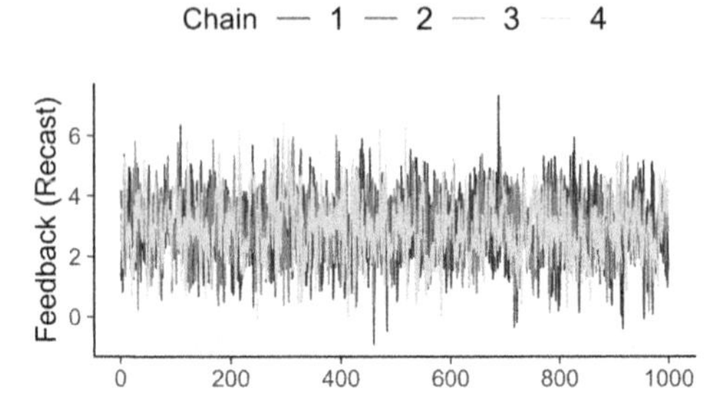

FIGURE 10.7 Trace Plot for Diagnostics in Bayesian Model

In Fig. 10.7, the x-axis represents the sampling process: by default, each chain in our model has 2,000 iterations (iter argument in brm()), but 1,000 of those are used as warmup (warmup argument in brm()).[15] As a result, we actually have 1,000 "steps" per chain in our random walk to find out what parameter values are more likely given the data being modeled. These 1,000 steps are represented by the x-axis in our figure. As you can see in the figure, all four chains have *converged* towards the same range of $\hat{\beta}$ values (on the y-axis).

The fact that you probably can't differentiate which line represents which chain in Fig. 10.7 is a good sign; it means that all four chains overlap, that is, they agree—it would help to use colors here (see line 23 in code block 56). If you go back to the estimates of fit_brm1, you will notice that $\hat{\beta} =$ 2.91 for FeedbackRecast—the value representing the peak of the posterior. On the y-axis in Fig. 10.7, you will notice that all four chains are clustered around the $\hat{\beta}$ value in question, that is, they all agree.

The code that generated Fig. 10.7 is provided in code block 56—this code block prepares our data not only for the trace plot shown in Fig. 10.7 but for our next plot (the code should be added to plotsBayesianModels.R). First, we extract the values from our posterior distributions (line 7)—run line 8 to see what that looks like. Line 11 prints the names of the parameters from our model (recall that this is a hierarchical model). Line 14 renames some of the parameters—those we're interested in plotting. Line 19 creates a vector to hold only the estimates of interest. Finally, we select a color scheme (line 22) and create our plot (lines 26–28)—you may want to run line 23 to produce a trace plot with color, which will help you see the different chains plotted.

Next, let's plot what's likely the most important figure when it comes to reporting your results: the model estimates. Overall, it's much easier to see a model's estimates in a figure than in a table. Fig. 10.8 should be familiar: we have done this before (e.g., Fig. 7.7), but now we're not plotting

R code

```
# You must run fit_brm1 before running this code block
library(bayesplot)
# library(extrafont)

# Prepare the data for plotting
# Extract posteriors from fit:
posterior1 = as.array(fit_brm1)
head(posterior1)

# Check all parameters in the model:
dimnames(posterior1)$parameters

# Rename the ones we'll use:
dimnames(posterior1)$parameters[c(1:3, 8)] = c("Intercept",
                                               "Feedback (Recast)",
                                               "Task (Task B)", "Sigma")

# Select which estimates we want to focus on:
estimates = c("Intercept", "Feedback (Recast)", "Task (Task B)", "Sigma")

# Select color scheme:
color_scheme_set("gray")      # To produce figures without color
# color_scheme_set("viridis") # Use this instead to see colors

# Create plot:
mcmc_trace(posterior1, pars = "Feedback (Recast)") +
  theme(text = element_text(family = "Arial"),
        legend.position = "top")

# ggsave(file = "figures/trace-fit_brm1.jpg", width = 4, height = 2.5, dpi = 1000)
```

CODE BLOCK 56 Creating a Trace Plot

point estimates and confidence intervals. We're instead plotting posterior distributions with credible intervals. This is the figure you want to have in your paper.

Fig. 10.8 is wider on purpose: because our intercept is so far from the other parameters, it would be difficult to see the intervals of our estimates if the figure were too narrow. That's the reason that you can't see the actual distributions for Task or Sigma (estimated variance) they're too narrow relative to the range of the *x*-axis.[16] As you can see, no posterior distribution includes zero. That means that zero is not a credible value for any of the parameters of interest here. Note that *even if* zero were included, we'd still need to inspect *where* in the posterior zero would be. If zero were a value at the tail of the distribution, it'd still not be very credible, since the most credible values are in the middle of our posteriors here—which follow a normal distribution.

Finally, Fig. 10.8 shows two intervals for each posterior (although you can only see both for the intercept in this case). The thicker line represents the

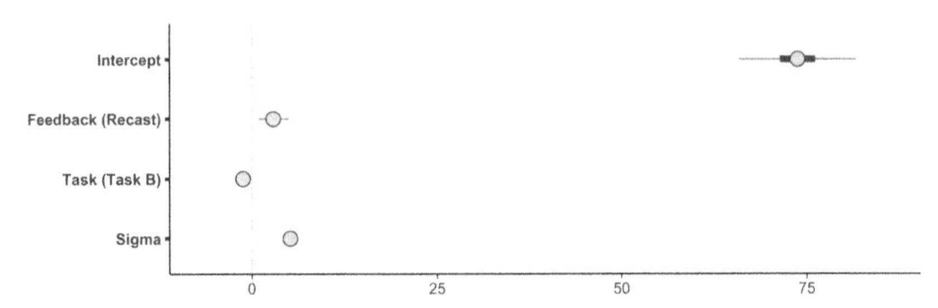

FIGURE 10.8 Plotting Model's Estimates (Posterior Distributions)

R code

```
1 # Plot estimates with HDIs:
2 mcmc_intervals(posterior1,
3               pars = estimates,
4               prob = 0.5,
5               prob_outer = 0.95) +
6   theme(text = element_text(family = "Arial"))
7
8 # ggsave(file = "figures/estimates-fit_brm1.jpg", width = 8, height = 2.5, dpi = 1000)
```

CODE BLOCK 57 Code for Plotting Model's Estimates (Fig. 10.8)

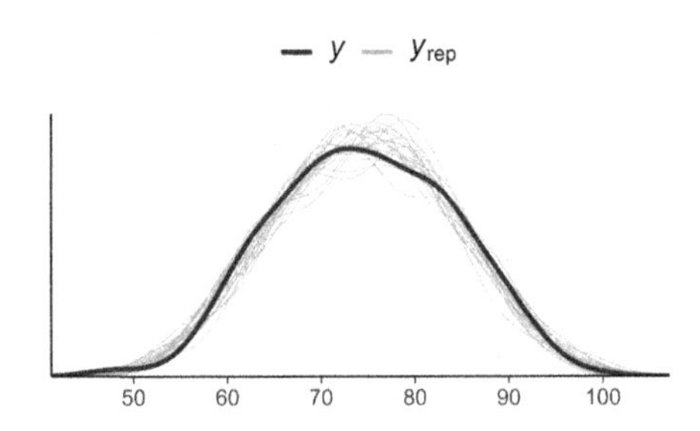

FIGURE 10.9 Posterior Predictive Check

50% HDI, while the thinner line represents the 95% HDI—use lines 4 and 5 in code block 57 if you wish to plot different intervals.

So far, we have checked some diagnostics ($\hat{R}$, $\hat{n}_{eff}$, trace plot), and we have plotted our model's estimates. Another quick step to check whether our model is appropriate is to perform a *posterior predictive check*. The intuition is simple: data generated from an appropriate model should look like the real data originally fed to our model.

Fig. 10.9 plots our actual data (thick black line), represented by y, against replicated data from our model, represented by y_{rep}—here, we are using 20 samples from our data. The x-axis represents predicted scores, our response variable. As you can see, the patterns from the predicted data match our actual data quite well. Although you don't need to include such a figure in your paper, it's important to perform predictive checks to ensure that our models are actually adequate given the data.

REPORTING RESULTS

A hierarchical linear model confirms that Feedback has a statistically credible effect on participants' scores in the experimental data under analysis ($\hat{\beta}_{\text{Recast}} = 2.91$, 95% HDI = [0.92, 4.96]). Task also has a statistically credible effect ($\hat{\beta}_{\text{TaskB}} = -1.18$, 95% HDI = [−2.25, −0.12])—see Fig. 10.8. Our model included by-participant random intercepts and slopes (for Task) as well as by-item random intercepts. Convergence was checked based on $\hat{R}$ and $\hat{n}_{eff}$. Posterior predictive checks confirm that the model generates simulated data that resemble actual data.

The earlier paragraph is likely more than you need: it's a comprehensive description of the results coming out of fit_brm1. Before reporting the results, you will naturally describe the methods used in your study. That's where you should clearly explain the structure of your model (e.g., which predictors you are including). You should also motivate your Bayesian analysis (§10.1) by highlighting the advantages of this particular framework relative to Frequentist inference. For more details on how to report Bayesian models, see Kruschke (2015, ch. 25).

Finally, you should carefully explain what a Bayesian estimate means, since your reader will probably not be used to Bayesian models. Fortunately, it's much more intuitive to understand Bayesian parameter estimates than Frequentist ones. Again: these estimates are simply telling us what the most credible effects of Feedback and Task are *given the data* being modeled.

10.4.2 Bayesian Model B: Relative Clauses with Prior Specifications

Our next model will be a hierarchical logistic regression, which will focus on our hypothetical relative clause study (chapter 7). This time, however, we will set the prior distributions ourselves.

Recall that Spanish has a tendency to favor high attachment, while English has a tendency to favor low attachment—you may want to briefly revisit chapter 4, especially (2), to review the hypothetical study in question. On top of that, we know given the literature that learners transfer their L1 patterns onto the L2. Thus, we have certain expectations regarding Spanish-speaking participants in our study. Here are three expectations we could consider (there are many more).

- English speakers should favor low attachment more than 50% of the time in the NoBreak condition (the "control" condition).
- In the Low condition, English speakers should favor low attachment even more.
- Spanish learners should favor low attachment *less* often than native speakers in the NoBreak condition.

The goal here is to use brms to run a hierarchical (and Bayesian) version of fit_glm4, first examined in chapter 7. Our model specification will be Low ~ Condition * Proficiency + (1 | ID) + (1 | Item). Recall that condition has three levels (High, Low, NoBreak) and that proficiency also has three levels (Adv, Int, Nat). As usual, we will choose our reference levels to be NoBreak for Condition and Nat for Proficiency—this is the more intuitive option here. Consequently, our intercept will represent the expected response (in log-odds) when Condition = NoBreak and Proficiency = Nat—this should be familiar, since we're basically repeating the rationale we used for fit_glm4 back in chapter 7.

The model specified here, which we'll call fit_brm2, will be used to predict the log-odds of choosing Low attachment, which we can map onto probabilities by using the inverse logit function discussed in chapter 7. Our response variable is binary, so we cannot assume that it follows a normal distribution (cf. fit_brm1). We use the logistic function ($logit^{-1}$) to map 0s and 1s from our response variable Low onto probabilities, and our probabilities will follow a Bernoulli distribution—*Bern* in 10.4. Nothing here is conceptually new (see chapter 7). What *is* new is that we can specify our priors for different parameters.

$$
\begin{aligned}
y_i &= Bern(\mu_i) \\
\mu_i &= logit^{-1}(\beta_0 + \alpha_{j[i]} + \alpha_{k[i]} + \beta_{hi}x_{i1} + \beta_{lo}x_{i1} + \ldots) \\
\beta_0 &\sim \mathcal{N}(1, 0.5) \\
\beta_{lo} &\sim \mathcal{N}(1, 0.5) \\
\beta_{int} &\sim \mathcal{N}(-1, 0.5) \\
\ldots &\sim \ldots
\end{aligned}
\tag{10.4}
$$

Look at the third and fourth lines in 10.4. We are essentially saying that we expect both β_0 and β_{lo} to follow a normal distribution with a positive mean ($\mu = 1$). What does that mean? For the intercept, it means that we

R code

```
source("dataPrepCatModels.R")

# Adjust reference levels:
rc = rc %>%
  mutate(Condition = relevel(Condition, ref = "NoBreak"),
         Proficiency = relevel(Proficiency, ref = "Nat"))

# Get priors:
get_prior(Low ~ Condition * Proficiency + (1 | ID) + (1 | Item), data = rc)

# Specify priors:
priors = c(prior(normal(1, 0.5), class = "Intercept", coef = ""),        # Nat, NoBreak
           prior(normal(1, 0.5), class = "b", coef = "ConditionLow"),     # Nat, Low
           prior(normal(-1, 0.5), class = "b", coef = "ProficiencyInt")) # Int, NoBreak

# Run model:
fit_brm2 = brm(Low ~ Condition * Proficiency +
                 (1 | ID) + (1 | Item),
               data = rc,
               family = bernoulli(),
               save_model = "fit_brm2.stan",
               seed = 6,
               prior = priors)

fit_brm2

# save(fit_brm2, file = "fit_brm2.RData")
# load("fit_brm2.RData")
```

CODE BLOCK 58 Specifying Priors and Running our Model

expect native speakers in the NoBreak condition to favor low attachment (positive estimates → preference for low attachment above 50%). For β_{lo}, it means that we also expect a positive estimate for the low condition—remember that our reference levels are the NoBreak condition and the Nat proficiency. In other words, we expect native speakers to favor low attachment *more* in the low attachment condition (relative to the NoBreak condition). Now look at the fifth line, which says we expect intermediate participants to have an effect that follows a normal distribution with a negative mean ($\mu = -1$). This means that we expect these participants to favor low attachment *less* relative to native speakers (in the NoBreak condition). These are the only prior distributions we will specify, but you could specify all the terms in the model (our model here has random intercepts as well as interaction terms).

Code block 58 shows how to specify priors in a Bayesian model run with brms. You should run line 9 to understand the structure of priors first. Lines 12–14 then specify prior distributions for the three parameters discussed earlier—the code should be self-explanatory.

The model in question, fit_brm2, is run in lines 17–23 (notice how we specify our priors in line 23). As with fit_brm1, you can print the results of the model by running line 25. Lines 27 and 28 simply show how to save the model (so you don't have to rerun it again in the future) and how to load it

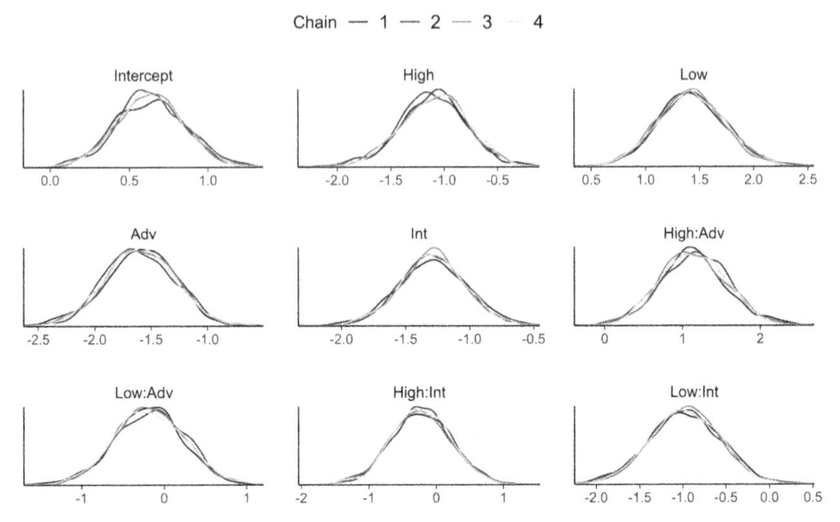

FIGURE 10.10 Posterior Distributions for all Parameters of Interest (Four Chains Each)

into RStudio once you've saved it—exactly the same code we used for fit_brm1. This time, however, code block 58 doesn't include an excerpt of the output of the model. That's because we will plot the estimates later.

Fig. 10.10 is a useful way to visualize our model's results.[17] First, we have nine panels (for the nine parameters of interest in our model). Second, we have the entire posterior distributions for each parameter, so we don't just see the mean of the distributions (which is what we'd see in the estimates column in our output). Third, for each parameter, we see the posterior distribution for each of the four chains in our simulation, which means we can also inspect if the chains converged (this is similar to inspecting a trace plot). We can clearly see that the chains converged, since all four lines for each parameter are showing us roughly the same posterior distribution of credible parameter values.

You may also want to zoom in on specific parameters to see their combined posterior distributions. Fig. 10.11 plots the posterior distribution of the intercept from fit_brm2 against the posterior distribution of $\hat{\beta}_{lo}$—this would be equivalent to a 2D version of what is shown in Fig. 10.6. The code used to generate both figures from fit_brm2 is shown in code block 59—Fig. 10.11 requires the hexbin package.

Finally, you should also inspect $\hat{R}$ and $\hat{n}_{eff}$ for fit_brm2, just as we did for fit_brm1. Likewise, you should run a posterior predictive check to make sure that the data predicted by the model mirrors actual data.[18]

The effects of the model in question are not too different from the estimates of fit_glm4. This is mostly because our priors are only mildly informative here

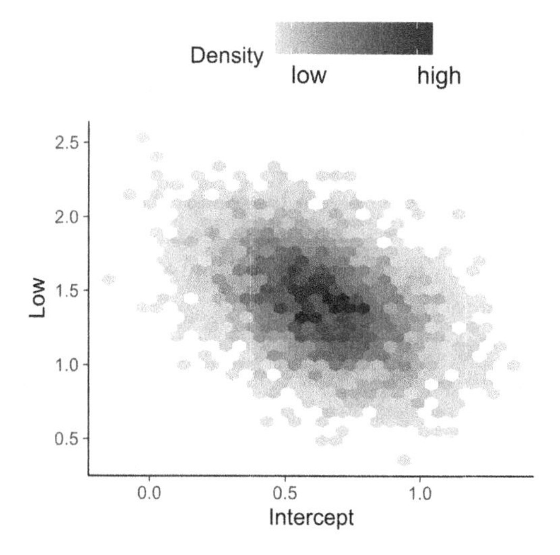

FIGURE 10.11 Posterior Distributions for Two Parameters

(the standard deviations of our priors are wide enough that the model can easily override our priors given enough evidence in the data)—you will have the opportunity to play around with priors in the exercises that follow. Of course, you should bear in mind that fit_brm2 is not only different from fit_glm4 because we're running Bayesian models here: it's also different because fit_brm2 is hierarchical and includes random intercepts for participants and items.

Last but not least, we can compare different Bayesian models much like we compared different Frequentist models back in chapters 6, 7, and 9. There are multiple ways to compare model fits (Bayesian or Frequentist). You may remember our discussion on AIC values from chapter 7. A similar metric can be used for Bayesian models, but it's called Widely Applicable Information Criteria (**WAIC**). For example, say you have two models, modelA and modelB, and you want to decide which one to report. You can run waic (modelA, modelB) (or run waic() on each model individually) to see WAIC values for both models. As with AIC values, the lower the WAIC value the better the fit. Another technique commonly used to assess a model's fit is Leave-one-out cross validation (**LOO**), which you can run using the loo() function (it works the same way as waic()). As with WAIC values, lower looic values in the output of loo() indicate a better fit.

10.5 Additional Readings on Bayesian Inference

This chapter was a brief introduction to Bayesian inference. We have barely scratched the surface, but you should be able to run and interpret your own models using R given what we discussed here. The main goal here was to get you started with the fundamental concepts and the code to implement Bayesian models through brms. Combined with what we discussed in previous chapters, you can now apply the code from this chapter to run linear, logistic, and ordinal hierarchical models using Bayesian estimation. You may also take a look at the tidybayes package (Kay 2020), which brings tidy data to Bayesian models (and also offers additional options for data visualization). Following are some reading suggestions should you wish to learn more about Bayesian statistics.

If you're interested in the fascinating history of Bayes's theorem, see McGrayne (2011). If you are interested in the intersection between philosophy of science and statistics and are wondering whether Bayesian inference is best characterized as inductive inference, see Gelman and Shalizi (2013).

R code

```
1  # You must run fit_brm2 before running this code block
2  # Prepare the data for plotting
3  # Extract posteriors from fit:
4  posterior2 = as.array(fit_brm2)
5  head(posterior2)
6
7  # Check all parameters in the model:
8  dimnames(posterior2)$parameters
9
10 # Rename the ones we'll use:
11 dimnames(posterior2)$parameters[1:9] = c("Intercept", "High", "Low",
12                                          "Adv", "Int", "High:Adv",
13                                          "Low:Adv", "High:Int", "Low:Int")
14
15 # Select which estimates we want to focus on:
16 estimates2 = c("Intercept", "High", "Low",
17                "Adv", "Int", "High:Adv",
18                "Low:Adv", "High:Int", "Low:Int")
19
20 # Select color scheme:
21 color_scheme_set("gray") # To produce figures without color
22 # color_scheme_set("viridis") # Use this instead to see colors
23
24 # Create plot: posterior distributions for all 9 parameters of interest
25 # Plot chains overalaid (not shown; see fit_brm2)
26 mcmc_dens_overlay(posterior2, pars = estimates2) +
27   theme(legend.position = "top",
28         text = element_text(family = "Arial"))
29
30 # ggsave(file = "figures/posteriors-brm2.jpg", width = 8, height = 5, dpi = 1000)
31
32 # Create plot: posterior distributions for intercept and Condition = Low
33 library(hexbin) # Install hexbin package first
34 mcmc_hex(posterior2, pars = c("Intercept", "Low")) +
35   theme(legend.position = "top",
36         text = element_text(family = "Arial"))
37
38 # ggsave(file = "figures/hex-brm2.jpg", width = 4, height = 4, dpi = 1000)
```

CODE BLOCK 59 Plotting Posterior Distributions (Figs. 10.10 and 10.11)

The top three books I would recommend on Bayesian data analysis are Gelman et al. (2014a), Kruschke (2015), and McElreath (2020). You may want to read Kruschke and Liddell (2018), an introductory and user-friendly article for newcomers. Then, if you decide you want to really explore Bayesian methods, start with Kruschke (2015) or McElreath (2020), which are the most user-friendly options of the three books.

Much like the present chapter, all three books suggested here focus on data-analytic applications, but you may also be interested in the applications of Bayesian models in cognition and brain function. In that case, see Chater et al. (2006), Tenenbaum et al. (2006), as well as Lee and Wagenmakers (2014).

Finally, if you'd like to see Bayesian data analysis applied to second language research, see Norouzian et al. (2018) and Garcia (2020). In Garcia (2020), I compare different hypotheses about language transfer by running statistical models using different sets of priors.

10.6 Summary

In this chapter, we ran two hierarchical models using Bayesian estimation. Bayesian inference offers many advantages (and some disadvantages) over Frequentist inference. Let's briefly review some of its advantages. First, Bayesian models are more intuitive to interpret, as they provide the probability of a parameter value given the data, and not the probability of the data given a parameter value. As a result, interpreting estimates and credible intervals is much more intuitive and natural than interpreting Frequentist confidence intervals. Second, Bayesian models allow us to incorporate our specialized knowledge into our statistical analyses in the form of prior distributions. This, in turn, provides the researcher with much more analytical power. Third, outputs of Bayesian models are considerably more comprehensive than those of Frequentist models, as we're given entire posterior distributions of credible parameter values. Fourth, Bayesian models allow for a higher degree of customization and complexity.

How about disadvantages? First, Bayesian models are computationally demanding. Running such models takes much longer than running their Frequentist counterparts. Second, Bayesian data analysis is still in its infancy in the field of second language research. Consequently, these methods will likely not be familiar to reviewers or readers. For example, many people may find it unsettling not to see *p*-values at all in your analysis, and you might have a hard time explaining why that's actually a good thing. Third, our field is mostly used to categorical answers to complex questions, so providing posterior distributions may backfire if not enough background is provided beforehand. Here's why: people tend to like *p*-values because they provide a clear and comforting categorical answer to our research questions (however complex such questions may be). Naturally, this is a simplistic understanding of statistical inference. By generating entire posterior distributions, we are instead estimating uncertainty, and our analysis becomes much more realistic by not implying that we can have categorical answers to all questions.

Let's briefly recap the main points of the chapter here.

- Any model can be run using Bayesian estimation. You only need to know which distribution to use for your model—unlike the models run in chapters 6–9, brms allows us to use a single function, namely, brm() to run all our models. When we ran our linear regression (fit_brm1), we specified family = gaussian(); for our logistic regression, we specified family = bernoulli(). If we wanted to run an ordinal regression, we'd specify family = cumulative("logit").
- Bayesian models provide posterior distributions of effect sizes, instead of the single-point estimates we get in Frequentist models.
- Estimates are credible parameter values given the data (cf. Frequentist inference).
- We can choose from a wide range of prior distributions to customize our models, even though we can also let brms choose its default distributions ("auto mode"). Our decisions regarding priors should be grounded on the literature, that is, on previous experiments that measure the effects of interest.
- If we use vague priors (i.e., if we let brms use default priors), then our results will be very similar to the results of a Frequentist model, since the posterior will be mostly determined by the data being modeled.
- You will often get warnings when running models using Stan. They are not necessarily bad; most of the time they're simply trying to help you with some information (see Appendix A.4). It's up to you to assess whether the warning is applicable or not. A comprehensive list of warnings is provided at https://mc-stan.org/misc/warnings.html.
- To check whether our model is appropriate, and that it has converged, we should run a posterior predictive check, and we should inspect $\hat{R}$ as well as the $\hat{n}_{eff}$ for each parameter being estimated.
- Because our estimates are given in distributions, plotting said estimates is probably the best way to present our results.

10.7 Exercises

Ex. 10.1 Basics

1. What are the main differences between Bayesian credible intervals (e.g., highest density intervals) and Frequentist confidence intervals?
2. Why do we need more than one chain in our sampling process?
3. Most of what we discussed in previous chapters can be naturally applied to Bayesian models. For example, you may want to rescale both predictor

variables in fit_brm1 before running the model. Why would that be unnecessary in this case?

4. Consider the following assumption: native speakers tend to have faster reaction times than non-native speakers. In fit_glm1, back in code block 34 (chapter 7), we modeled L2er ~ RT. Our RT estimate, $\hat{\beta} = 1.38$, represented the change in log-odds of being an L2er for every additional unit of reaction time, that is, one second. Simply put, the positive effect confirms that L2ers are on average slower relative to native speakers. Now suppose that we were to run a similar experiment. This time, we could use our estimate to incorporate mildly informed priors into our model—after all, we already know what to expect given the literature and our own estimate. We would be telling our model that we expect our estimate for RT to follow a specific distribution. Which of the following Gaussian distributions would be a reasonable prior for RT in our new model?

a) RT$\sim \mathcal{N}(-1.38, 0.01)$ (Normal distribution, $\mu = -1.38$ and $\sigma = 0.01$)
b) RT$\sim \mathcal{N}(1.38, 0.01)$ (Normal distribution, $\mu = 1.38$ and $\sigma = 0.01$)
c) RT$\sim \mathcal{N}(-1.38, 1)$ (Normal distribution, $\mu = -1.38$ and $\sigma = 1$)
d) RT$\sim \mathcal{N}(1.38, 1)$ (Normal distribution, $\mu = 1.38$ and $\sigma = 1$)

Ex. 10.2 Bayesian Ordinal Model

1. Run the equivalent to fit_clm1 using brms—remember to relevel Condition such that NoBreak is the reference level. You will need to set the family argument correctly—see §10.6. Call this model fit_brm4_naive. Inspect the output—both $\hat{\tau}$ and $\hat{\beta}$ estimates. How do the estimates compare with the Frequentist version of the model? Interpret the results.
2. Generate a plot for the estimates of the model you just ran (use Fig. 10.8 and associated code as a template).
3. Now run a more informed version of the same model—call it fit_brm4_inf1. This time, add the following priors (use code block 58 as a guide):

   ConditionLow $\sim \mathcal{N}(-1, 0.1)$
   ConditionHigh $\sim \mathcal{N}(1, 0.1)$

   Once you've run the model, run another model, fit_brm4_inf2. This time, add these priors:

   ConditionLow $\sim \mathcal{N}(-1, 0.01)$
   ConditionHigh $\sim \mathcal{N}(1, 0.01)$

   Note that, since we're only analyzing Spanish speakers, the prior for ConditionHigh follows a normal distribution with a *positive* mean—after all, Spanish is known to favor high attachment. The opposite is true of ConditionLow. How do the estimates for the two informed models compare to those from fit_brm4_naive? Why do you think the estimates for ConditionLow and ConditionHigh in fit_brm4_inf2 are exactly |1.00|? Run plot() on all three models to inspect their posterior distributions (and trace plots).

4. Now that you have three models, compare them using the loo() function. Simply run loo(fit_brm4_naive, fit_brm4_inf1, fit_brm4_inf2)—this can take a few seconds to run. Three outputs will be printed, one for each model—a model comparison is also printed in your console. Which model has the lowest looic value (and therefore the best fit) of the three? As you scroll down the output of loo(), examine the model comparison and focus on the elpd_diff column. The model at the top will be fit_brm4_inf2. Both elpd_diff and se_diff will be zero, since here we're comparing the model with itself. Now look at the other two models, which are compared to fit_brm4_inf2. The negative elpd_diff values indicate worse fits relative to fit_brm4_inf2. Crucially, the standard error of the difference (se_diff) is more than twice as large as the value for elpd_diff for both models being compared to fit_brm4_inf2. Therefore, these models are all statistically different.
5. Considering your answers to the previous questions, which of the three models would you choose to report? Why?

Notes

1. Named after English mathematician Thomas Bayes (1702–1761; Bayes 1763). Across the English channel, French scholar Pierre-Simon Laplace independently developed and further explored the probability principles in question—see Dale (1982) and McGrayne (2011) for a comprehensive historical account.
2. A rephrasing of Laplace's principle. Sagan's sentence is indeed a characterization of Bayesian inference.
3. Note that the idea that what is more probable is what happens more often is much older than Frequentist statistics.
4. For mathematical details, see Kruschke (2015, ch. 6). Simply put, a beta distribution is a continuous probability distribution—which means it is constrained to the interval [0, 1]. Notice that the prior distribution in the figure is not symmetrical (cf. Gaussian distribution). The shape of a beta distribution relies on two parameters, typically denoted as α and β—much like the shape of a Gaussian distribution also relies on two parameters, μ and σ. The shape of the prior distribution here is defined as $\alpha = 3$, $\beta = 10$.
5. We can also calculate the *equal-tailed interval*, or ETI. See discussion in Kruschke (2015, pp. 242–243).
6. Note that 95% is an arbitrary number—as arbitrary as 89%. Here we will keep using 95% as our standard interval, but always remember that there's nothing special about this number (other than most people seem to like it).
7. Remember that a probability of 0.5 is equivalent to 0 log-odds (Table 7.1) in a logistic regression.

8. A popular alternative is to use JAGS (*Just Another Gibbs Sampler*). For our purposes here, JAGS and Stan are very similar, but see Kruschke (2015, ch. 14) for a discussion on the differences between these two options.
9. Unlike robust linear regressions, which assume a t-distribution—see, for example, Gelman et al. (2014a, ch. 17).
10. We will not use 3D figures, though, since they can only display the joint distributions of two parameters at once. If we have a model with n predictors (where $n > 2$), we can't actually picture what their joint distribution looks like—we can, of course, look at the their distributions individually.
11. You may get a warning message after running Bayesian models. See Appendix A.4 if that happens.
12. You may remember from chapter 7 that we specified the family of our model for logistic regressions, so this is not exactly new. To run a robust linear regression, specify `family = student()`—see Kruschke (2015, §17.2) for the implementation of such a model. If you're not familiar with robust models, also see Gelman et al. (2014a, ch. 17).
13. Recall that our intercept represents the predicted score of a participant when `Feedback = Explicit feedback` and `Task = TaskA`. This is the same interpretation as before.
14. You will often get warnings and/or see that something's off by inspecting the output of the model, but visualizing the chains is perhaps the most intuitive way to check that they have converged.
15. You can naturally change these default values by adding these two arguments to `brm()` when you run the model. The more iterations our chains have, the more accurate our posterior estimates will be—and the longer it will take for the model to run.
16. You could remove the intercept from the figure by adjusting the code shown in code block 56, line 19: simply remove the intercept from the vector and rerun the code, and then run code block 57.
17. Your figure will look slightly different.
18. Note that this time the response variable is binary, so instead of the normal distribution shown in Fig. 10.9, you should go for a different type of figure. For logistic regressions, bars would be more useful. Simply run `pp_check(fit_brm2, type = "bars", nsamples = 20)` and see what happens. You can run this multiple times to see how the figure changes with different samples.

11

FINAL REMARKS

The two main goals of this book were to emphasize the importance of data visualization in second language research and to promote full-fledged statistical models to analyze data. I hope to have convinced you of the numerous advantages of visualizing patterns *before* statistically analyzing them and of analyzing such patterns using hierarchical models, which take into account individual- and group-level variation when estimating group-level coefficients. More comprehensive quantitative methods are vital in our field, given that second language studies typically employ a dangerously narrow range of statistical techniques to analyze data—as mentioned in chapter 1.

In addition to these two points, I hope you are now convinced that Bayesian data analysis is not only more powerful and customizable but also more intuitive in many respects. Perhaps more importantly, Bayesian models allow us to incorporate our previous knowledge into our statistical analysis. Second language research can certainly benefit from using priors that mirror (i) previous research findings or (ii) L1 grammars in the context of language transfer (e.g., Garcia 2020).

Throughout this book, we examined several code blocks in R. The fact that R is a powerful language designed specifically for statistical computing allows you and me to easily reproduce the code blocks discussed earlier. And because R is open-source and so widely used these days, we have literally thousands of packages to choose from. Packages like tidyverse and brms allow more and more people to accomplish complex tasks by running simple lines of code.

We also saw in different chapters how files can be organized using R Projects. We saw that by using the source() function we can easily keep different components of our projects separate, which in turn means better organization overall. Finally, we briefly discussed RData files, which offer a great option to compress and save multiple objects into a single file.

The range of topics introduced in this book makes it impossible for us to discuss important details. For that reason, numerous reading suggestions were made in different chapters. The idea, of course, was to give you enough tools to carry out your own analyses but also to look for more information in the literature when needed.

The end.

Appendix A

Troubleshooting

A.1 Versions of R and RStudio

As noted in chapter 2, the code in this book was last tested using R version 4.0.2 (2020-06-22)—"Taking Off Again" and RStudio Version 1.3.1073 (Mac OS). Your console will tell you the version of R you have installed every time you open RStudio. To find out which version of RStudio you have, go to Help ≻ About RStudio. Needless to say, both R and RStudio change with time, and so do packages. As a result, new functions may be introduced or changed, and old functions may become obsolete (deprecated). As you run the code in this book, you might come across different warning messages depending on the version of R and RStudio you use. These messages are often self-explanatory, but you can always google them to learn more.

A.2 Different Packages, Same Function Names

Because R has so many packages, and because each package has several functions in it, it's not surprising that certain function names appear in more than one package. For example, in the package tidyverse, more specifically in the package dplyr, there's a function called select(), which we use to select the columns that we want from a data frame or tibble. It turns out that another package we have used, arm, also has a function called select(). In these situations, where there's ambiguity, R will choose one of the two functions unless you are explicit about which function you mean.

Try doing this: load tidyverse and then load arm. Different messages will be printed in your console, one of which says "The following object is masked from 'package:dplyr': select". This is R telling you that if you run select(), it will assume you mean the arm version, not the dplyr version, which is now "masked".

Therefore, if you have both packages loaded and plan to use select(), make sure you are explicit about it by typing dplyr::select()—you can find this in code block 34. Otherwise, you will get an error, since you'll be using the wrong function without realizing what the problem is.

A.3 Errors

When you start using R, error messages will be inevitable. The good news is that many errors are very easy to fix. First, you should check the versions of R and RStudio that you have installed (see §A.1 or §2.2.1). If your computer is too old, your operating system will likely not be up-to-date, which in turn may affect which versions of R and RStudio you can install. That being said, unless your computer is a decade old, you shouldn't have any problems. In some cases, an error will go away once you restart RStudio or your computer—updating packages may also solve the issue, in which case R will give you instructions. As already mentioned, all the code in this book has been tested on different operating systems, but each system is different, and you might come across errors.

It's important to notice that R is case-sensitive, so spelling matters a lot. For example, you may be running Head() when the function is actually head()—this will generate the error could not find function "Head". Or you may be running library(1me4) when it's called library(lme4)—this will generate the error unexpected symbol ... . Another common issue is trying to load a package that hasn't been installed yet—which will generate the error there is not package called ... . All these scenarios will throw an error on your screen, but all are easy to fix.

One common error you may come across is: Error in ...: object 'X' not found—replace X with any given variable. This happens when you are calling an object that is not present or loaded in your environment. For example, you try running the code to create a figure for the estimates of a given model, but you haven't run the actual model yet. Suppose you run some models, which are now objects in your environment. You then restart RStudio and try running some code that attempts to create a figure based on the model. You will get an error, since the models are no longer in RStudio's "memory" (i.e., its environment). You need to run the models—or source() the script that runs the models.

Another common error you may get is: Error in percent_format(): could not find function "percent_format". This is simply telling you that R couldn't find percent_format (used to add percentage points to axes in different plots). If you get this message, it's because you haven't loaded the scales package, which is where percent_format is located.

You may come across more specific or difficult errors to fix, of course. First, read the error message carefully—bear in mind that your error may be the result of some issue in your system, not in R *per se*. Second, check for typos in your code. Third, google the error message you got. Online forums will almost always help you figure it out—the R community is huge (and helpful!).

A.4 Warnings

As you run the code blocks in this book, you may come across warning messages. Unlike error messages, warning messages will not interrupt any processes—so they're usually not cause for alarm. If you run a function that is a little outdated, you may get a warning message telling you that a more updated version of the function is available and should be used instead. For example, if you run `stat_summary(fun.y = mean, geom = "point")`, you will get a warning message: `Warning message: 'fun.y' is deprecated. Use 'fun' instead`. Here, the message is self-explanatory, so the solution is obvious.

In chapter 10, you may run into warning messages such as "`There were 2 divergent transitions after warmup` [...]". This is merely telling you that you should visually inspect your model to check whether the chains look healthy enough (e.g., via a trace plot).

A.5 Plots

At first, your plots may look slightly different from the plots shown in this book, especially when figures have multiple facets—even though you are using the exact same code. But the difference is not actually real: once you save your plots using the `ggsave()` function, your figures will look exactly the same as mine—provided that you use the exact same code (including the specifications provided for `ggsave()`).

These apparent differences occur because when you generate a figure in RStudio, the preview of said figure will be adjusted to your screen size. A full HD laptop screen has much less room than a 4K external monitor. RStudio will resize the preview of your plot to fit the space available. As a result, if your screen is not 4K, you may see overlapping labels, points that are too large, and so on. You can resize the plot window to improve your preview, click on the Zoom button, or make changes to the code, but the bottom line is that once you save the figure, it will look perfectly fine, so this is not a problem.

Appendix B

RStudio Shortcuts

TABLE B.1 Keyboard Shortcuts in RStudio (Cmd on Mac = Ctrl on Linux or Windows)

Cmd + Shift + N	New script
Cmd + Enter	Run current line of code
Cmd + S	Save script
Cmd + Shift + C	Comment out a line of code (add "#")
Cmd + I	Reindent lines
Cmd + Shift + M	Add %>% ("pipe")
Cmd + ,	Open preferences (*only on Mac OS*)

Appendix C

Symbols and Acronyms

TABLE C.1 Main Symbols and Acronyms Used in the Book

N	Population size
μ	Population mean
σ	Population standard deviation
n	Sample size
$\bar{x}$	Sample mean
s	Sample standard deviation
H_0	Null hypothesis
SE	Standard Error
CI	Confidence Interval
β	Estimate in statistical models
α	Significance level in hypothesis testing
	Random intercept in hierarchical models
γ	Random slope in hierarchical models
ϵ	Error term
$\hat{e}$	Estimated error term
$\hat{y}$	Predicted response
$\mathcal{N}$	Normal (Gaussian) distribution
$\mathcal{U}$	Uniform distribution
HDI	Highest density interval
$P(A \mid B)$	Probability of A given B

Appendix D

Files Used in This Book

The structure shown in Fig. D.1 illustrates how the files are organized in different directories. Naturally, you may wish to organize the files differently: the structure used in this book is merely a suggestion to keep files in different directories and to use R Projects to manage the different components of the topics examined throughout the book. In a realistic scenario, you could have one directory and one R Project for each research project you have.

By adopting R Project files and using a single directory for all the files in your project, you can refer to said files *locally*, that is, you won't need to use full paths to your files. Of course, you can always import a file that is *not* in your current directory by using complete paths—or you can create copies of the file and have it in different directories if you plan to use the same data file across different projects. This latter option is in fact what you see in this book: `feedbackData.csv`, for example, is located in multiple directories (`plots`, `Frequentist`, `Bayesian` in the figure). These duplicated files (both scripts and data files) ensure that we have all the files we need within each working directory.

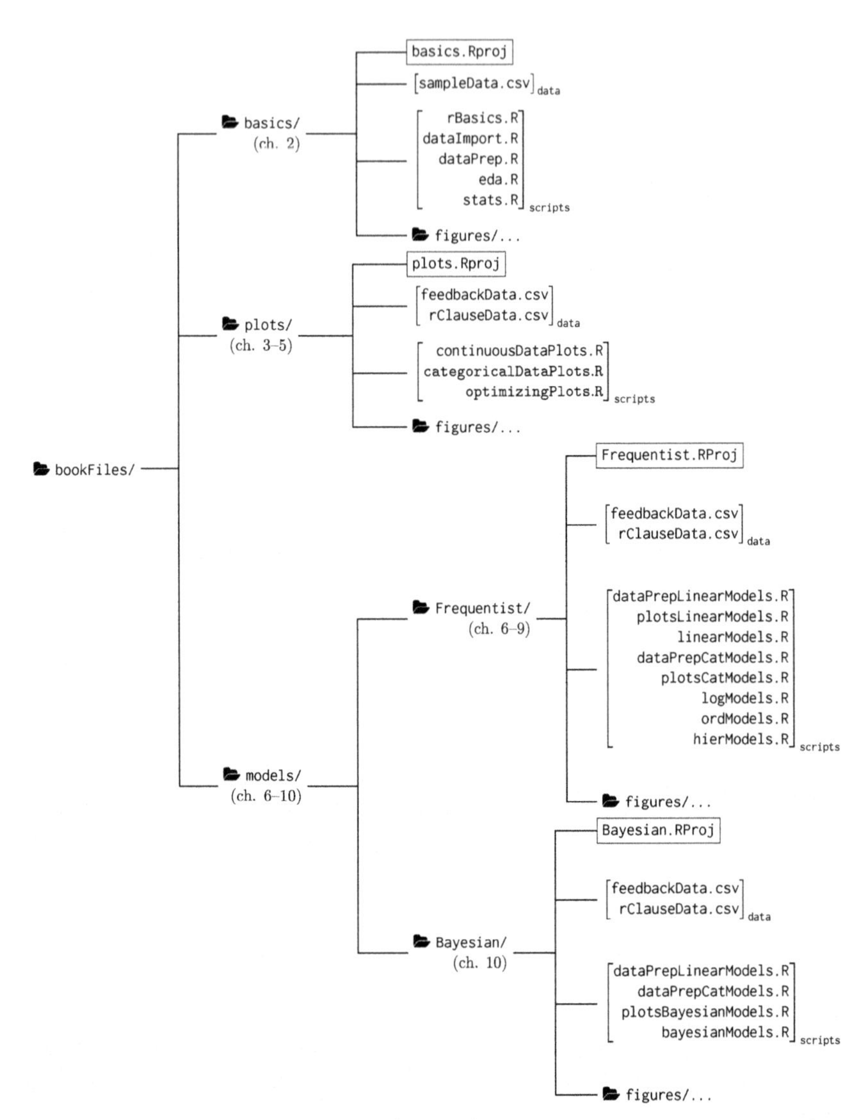

FIGURE D.1 File Structure Used in This Book

TABLE D.1 List of all Scripts and their Respective Code Blocks

Directory	*Script*	*Code block(s)*
basics/	rBasics.R	1, 2, 3
	dataImport.R	4
	dataPrep.R	5, 6, 7, 8
	eda.R	9
	stats.R	10
plots/	continuousDataPlots.R	11–19
	categoricalDataPlots.R	20, 21, 22
	optimizingPlots.R	23, 24
Frequentist/	dataPrepLinearModels.R	11 (updated in 47)
	plotsLinearModels.R	30, 31, 48, 53, 54
	linearModels.R	25–29, 32
	dataPrepCatModels.R	33
	plotsCatModels.R	37, 40, 42, 46
	logModels.R	34, 35, 36, 38, 39, 41
	ordModels.R	43, 44, 45, 60
	hierModels.R	49, 50, 51, 52
Bayesian/	dataPrepLinearModels.R	47
	dataPrepCatModels.R	33
	plotsBayesianModels.R	56, 57, 59
	bayesianModels.R	55, 58

Code block 60 is located in Appendix F

Appendix E

Contrast Coding

Here you can see how dummy coding works for a factor (L1) with three levels (German, Italian, Japanese). The factor in question, from feedback, is ordered alphabetically by R, which means German will be our reference level (i.e., our intercept) in a typical model—unless, of course, we manually change it using relevel(). Because we have three levels, we only need two columns in our contrast coding in Table E.1 (compare with Table 6.1 in chapter 6).

TABLE E.1 Example of Contrast Coding for More Than Two Levels

L1		*Italian*	*Japanese*
German		0	0
German		0	0
Italian		1	0
Japanese	→	0	1
Italian		1	0
Japanese		0	1
German		0	0
...		...	...

Appendix F

Models and Nested Data

In chapter 8 (code block 42), we run an ordinal regression on a subset of rc: we only look at Spanish speakers. We could, of course, run another model that only looks at English speakers, which would give us two separate models. However, dplyr (inside tidyverse) has yet another great function that can help us here: nest_by()—which could be added to Table 2.3 back in chapter 2.

The function in question separates the data into different subsets, which can then be fed to our statistical model. Simply put, it creates a nested tibble, such that a "cell" in your tibble can contain another tibble instead of a single value. By using summarize() and tidy()[1] (from the broom package), we can then extract estimates, standard errors, and so on. The whole process is shown in code block 60—lines 9–15. First, we nest_by() our variable of interest, L1. Next, in line 10, we create a new column in our tibble that will hold a model (note that the model fit is run inside list() and that data = data, not rc). We then summarize everything in line 11 and calculate confidence intervals in lines 12–13. Line 14 only selects beta values (as opposed to alpha values, which stand for intercepts or thresholds in our model). Finally, line 15 gets rid of some columns—you can comment the line out if you want all the columns, which will include z-values, p-values, and the coefficient type ($\hat{\beta}$ or $\hat{\tau}$).

The result of our code, shown in lines 20–25, displays estimates and confidence intervals for ConditionHigh and ConditionLow for both L1 groups: it's a very intuitive table that allows us to directly compare statistical models applied to different subsets of data. We could now use this table to create an informative figure comparing the groups.

R code

```
source("dataPrepCatModels.R")
library(tidyverse)
library(ordinal)

rc = rc %>%
  mutate(Condition = relevel(factor(Condition), ref = "NoBreak"))

# Create table with estimates and more: two models
rc %>% nest_by(L1) %>%
  mutate(model = list(clm(Certainty3 ~ Condition, data = data))) %>%
  summarize(broom::tidy(model)) %>%
  mutate(lowerCI = estimate - 1.96 * std.error,
         upperCI = estimate + 1.96 * std.error) %>%
  filter(coefficient_type == "beta") %>%
  dplyr::select(-c(statistic, p.value, coefficient_type))

# Output:
# A tibble: 4 x 6
# Groups:   L1 [2]
#   L1      term            estimate std.error lowerCI upperCI
#   <fct>   <chr>              <dbl>     <dbl>   <dbl>   <dbl>
# 1 English ConditionHigh     -0.631     0.264  -1.15   -0.114
# 2 English ConditionLow       0.730     0.270   0.201   1.26
# 3 Spanish ConditionHigh      1.08      0.190   0.704   1.45
# 4 Spanish ConditionLow      -0.877     0.192  -1.25   -0.501
```

CODE BLOCK 60 Running Multiple Models and Extracting Coefficients

Note

1. This function can also be used to display the results of a model, much like summary() and display(), that is, broom::tidy(fit). It's more minimalistic and organized than summary() but less minimalistic than display().

GLOSSARY

ANOVA *ANalysis Of VAriance*. Collection of models used to analyze the difference in means among groups in a sample. As in *t*-tests, the response/dependent variable under analysis must be continuous. p. 3

contrast coding When dealing with a categorical explanatory/predictor variable (i.e., a factor), R will choose one of the levels of the factor to represent its reference level. The remaining level(s) will be dichotomized. For example, in the case of Feedback, which has two levels, one level (Explicit correction) is chosen (alphabetically) as the reference level—this level will be represented by the intercept in a regression model. The remaining level, Recast, is recoded as 0 or 1. Thus, whenever we want to "activate" Explicit correction, we assign 0 to Recast. When we dummy code a factor with n levels, we will end up with $n - 1$ dichotomous (0/1) columns. See Table 6.1. p. 121 and p. 254

function Object that performs a task for us. Functions are like recipes to optimize different processes: we can either use existing functions (created by other people) or create our own functions, which will do exactly what we need. Functions almost always expect certain *arguments*. Understanding how to use a function involves understanding not only when to use it but also which arguments it requires (and which are optional). We will see numerous examples throughout the book. p. xxi (preface) and p. 15.

homoscedasticity The assumption that the error around our variable is constant across all values of the predictor variable. Taking Fig. 6.1 as an example, if our data is homoscedastic, then the distance from data points to the fitted line is constant across the x-axis, that is, ε (estimated as $\hat{e}$) is assumed to be constant. If our data doesn't meet this assumption, we say that we have *heteroscedastic* data. p. 114

leave-one-out cross-validation (LOO) Information criterion commonly used to assess a model's fit (mainly in relation to other models). The idea is to capture the out-of-sample prediction error by (i) training a given model on a dataset that contains all but one item (or set) and (ii) evaluating its accuracy at predicting said item (or set). p. 233

Markov Chain Monte Carlo (MCMC) Techniques used to fit Bayesian models. In a nutshell, MCMC methods draw samples from the posterior distribution instead of attempting to compute the distribution directly. The concept behind MCMC can be described as a random walk through the space of possible parameter values. Different MCMC algorithms can be used for this random walk, for example, Metropolis-Hastings, Gibbs, Hamiltonian. A Markov chain is a stochastic model where the probability of an event only depends on the previous state. Metaphorically, the next step we take in our random walk only depends on our previous step. Finally, "Monte Carlo" refers to the famous Casino in Monaco. p. 218

null hypothesis (H_0) The hypothesis that there is no significant difference between two groups. More generally, the null hypothesis assumes that there is no effect of a given variable. p. 7

***p*-hacking** Conscious or unconscious data manipulation to achieve a statistically significant result. See Nuzzo (2014). p. 8

pipe (%>%) Operator that takes the output of a statement in R and makes it the input of the next statement (see §2.5). p. 38

$\hat{R}$ Also known as the Gelman-Rubin convergence diagnostic. A metric used to monitor the convergence of the chains of a given model. $\hat{R}$ is a measure that compares the variance of the simulations from each chain with the variance of all the chains mixed together. If all chains are at equilibrium, $\hat{R}$ should equal 1. See Brooks et al. (2011). p. 255

shrinkage The reduction of variance in the estimators relative to the data being modeled. For instance, in hierarchical models where by-participant effects are taken into account, the individual-level predictions are shifted (i.e., shrunken) towards the overall mean. See Fig. 9.4. p. 201

slice notation The use of square brackets to access elements in a data structure, for example, vectors, lists, and data frames. For example, in a vector called `A`, we can run `A[2]` to pick the second element in `A`. In a data frame called `B`, we can run `B[3,5]` to pick the element found in the third row, fifth column; `B[,5]` picks all the rows in the fifth column; and `B[3,]` picks all the columns but only the third row. Because data frames have two dimensions, we use a comma to separate rows from columns when using slice notation. p. 25 and p. 29

string Sequence of characters. Each of the following is a string in R: "`hello`", "`I like chocolate`", "384", "a", "!!", "`dogs and cats?`". Notice that *any* sequence of characters will be a string as long as you have quotes around it. p. xxi (preface) and p. 24

t-test Perhaps the most popular statistical test used in *null hypothesis significance testing* (NHST). *t*-tests are used to determine whether there's a statistical difference in means between two groups. Unlike ANOVAs, *t*-tests can't handle more than two groups. As in ANOVAs, *t*-tests require the response/dependent variable to be continuous. p. 3

tibble Data structure that is more efficient than a data frame. Unlike data frames, tibbles require a package to be used (tibble, which comes with tidyverse). For more information (and examples), check the documentation at https://tibble.tidyverse.org. p. 31 and p. 40

tidy data "Tidy data" (Wickham et al. 2014) is data with a specific format that optimizes both exploratory data analysis and statistical analysis. A data frame or tibble is tidy if every variable goes in a column and every column is a variable (see more at https://www.tidyverse.org/packages/). p. 37

Tukey HSD Tukey's honestly significant difference generates multiple pairwise comparisons while controlling for Type I error, which would be a problem if we ran multiple *t*-tests. p. 53

Type I error Also known as a false positive conclusion, Type I error is the rejection of the null hypothesis when it is, in fact, *true*. p. 4

Type II error Also known as a false negative conclusion, Type II error is the failure to reject the null hypothesis when it is, in fact, *false*. p. 8

variable Object used to store information to be referenced later. For example, if we establish that x = 10, then every time we use x we will be essentially using 10, that is, the value of the variable in question. p. xxi (preface) and p. 22

vector Basic data structure in R. A vector can hold multiple elements as long as all its elements belong to the same class. For example, we can have a vector with numbers, a vector with characters, and so on. In a typical dataset, each column is a vector. p. 24

widely applicable information criterion (WAIC) Watanabe (2010), also known as *Watanabe Akaike information criterion*. A cross-validation method to assess the fit of a (Bayesian) model. It is calculated by averaging the log-likelihood over the posterior distribution taking into account individual data points. See McElreath (2020, p. 191) for a discussion on the differences among WAIC, AIC, BIC, and DIC. For advantages of WAIC over DIC, see Gelman et al. (2014b). p. 233

working directory The directory (or folder) assumed by R to be the location of your files. As a result, if you need to reference a file in your working directory, you can just tell R the name of the file and R will know where it is. You can find out what your working directory is by running the getwd() function, and you can change it with the setwd() function. You can also reference files that are *not* in your working directory, but doing so will require that you fully specify the path to the file in question. For example, if a file called myFile.csv is located in your (Mac OS) desktop, and your working directory is *not* currently set to your desktop, you will need to tell R that the file you want is "~/Desktop/myFile.csv". If the file were in your working directory, you would instead tell R that the file is "myFile.csv". p. 19

REFERENCES

Agresti, A. (2002). *Categorical data analysis*. John Wiley & Sons, Hoboken, New Jersey.

Agresti, A. (2010). *Analysis of ordinal categorical data*. John Wiley & Sons, Hoboken, New Jersey.

Akaike, H. (1974). A new look at the statistical model identification. *IEEE Transactions on Automatic Control*, 19(6):716–723.

Baayen, R. H. (2008). *Analyzing linguistic data: A practical introduction to statistics using R*. Cambridge University Press, New York.

Bache, S. M. and Wickham, H. (2014). *magrittr: A forward-pipe operator for R*. R package version 1.5.

Barr, D., Levy, R., Scheepers, C., and Tily, H. (2013). Random effects structure for confirmatory hypothesis testing: Keep it maximal. *Journal of Memory and Language*, 68(3):255–278.

Bartoń, K. (2020). *MuMIn: Multi-model inference*. R package version 1.43.17.

Bates, D., Kliegl, R., Vasishth, S., and Baayen, H. (2015a). Parsimonious mixed models. arXiv:1506.04967.

Bates, D., Mächler, M., Bolker, B., and Walker, S. (2015b). Fitting linear mixed-effects models using lme4. *Journal of Statistical Software*, 67(1):1–48.

Bayes, T. (1763). LII. An essay towards solving a problem in the doctrine of chances. By the late Rev. Mr. Bayes, F.R.S. communicated by Mr. Price, in a letter to John Canton, A.M.F.R.S. a letter from the late Reverend Mr. Thomas Bayes, F.R.S., to John Canton, M.A. and F.R.S. Author(s): Mr. Bayes and Mr. Price. *Philosophical Transactions (1683–1775)*, 53:370–418.

Brooks, S., Gelman, A., Jones, G. L., and Meng, X.-L. (2011). *Handbook of Markov Chain Monte Carlo*. Chapman and Hall/CRC, Boca Raton.

Bürkner, P.-C. (2018). Advanced Bayesian multilevel modeling with the R package brms. *The R Journal*, 10(1):395–411.

Campbell, J. P. (1982). Editorial: Some remarks from the outgoing editor. *Journal of Applied Psychology*, 67(6):691–700.

Carpenter, B., Gelman, A., Hoffman, M., Lee, D., Goodrich, B., Betancourt, M., Brubaker, M., Guo, J., Li, P., and Riddell, A. (2017). Stan: A probabilistic programming language. *Journal of Statistical Software, Articles*, 76(1):1–32.

Chater, N., Tenenbaum, J. B., and Yuille, A. (2006). Probabilistic models of cognition: Conceptual foundations. *Trends in Cognitive Sciences*, 10(7):287–344.

Christensen, R. H. B. (2019). *ordinal: Regression models for ordinal data*. R package version 2019.12-10.

Cuetos, F. and Mitchell, D. C. (1988). Cross-linguistic differences in parsing: Restrictions on the use of the late closure strategy in Spanish. *Cognition*, 30(1):73–105.

Cunnings, I. (2012). An overview of mixed-effects statistical models for second language researchers. *Second Language Research*, 28(3):369–382.

Dale, A. I. (1982). Bayes or Laplace? An examination of the origin and early applications of Bayes' theorem. *Archive for History of Exact Sciences*, 27:23–47.

Dowle, M. and Srinivasan, A. (2019). *data.table: Extension of 'data.frame'*. R package version 1.12.8.

Efron, B. and Morris, C. (1977). Stein's paradox in statistics. *Scientific American*, 236 (5):119–127.

Fairbanks, M. (2020). *tidytable: Tidy Interface to 'data.table'*. R package version 0.5.5.

Fernández, E. M. (2002). Relative clause attachment in bilinguals and monolinguals. In *Advances in Psychology*, volume 134, pages 187–215. Elsevier, Amsterdam.

Fodor, J. D. (2002). Prosodic disambiguation in silent reading. In Hirotani, M., editor, *Proceedings of the 32nd annual meeting of the North East linguistic society*. GLSA Publications, Amherst, MA.

Fullerton, A. S. and Xu, J. (2016). *Ordered regression models: Parallel, partial, and non-parallel alternatives*. Chapman & Hall/CRC, Boca Raton.

Garcia, G. D. (2014). *Portuguese Stress Lexicon. Comprehensive list of non-verbs in Portuguese*. Available at http://guilhermegarcia.github.io/psl.html.

Garcia, G. D. (2017). Weight gradience and stress in Portuguese. *Phonology*, 34(1):41–79. Project materials available at http://guilhermegarcia.github.io/garciaphon2017.html.

Garcia, G. D. (2020). Language transfer and positional bias in English stress. *Second Language Research*, 36(4):445–474.

Gelman, A. (2008a). Objections to Bayesian statistics. *Bayesian Analysis*, 3(3):445–449.

Gelman, A. (2008b). Scaling regression inputs by dividing by two standard deviations. *Statistics in Medicine*, 27(15):2865–2873.

Gelman, A., Carlin, J. B., Stern, H. S., Dunson, D. B., Vehtari, A., and Rubin, D. B. (2014a). *Bayesian data analysis*, volume 2. Chapman & Hall/CRC, Boca Raton, 3rd edition.

Gelman, A. and Hill, J. (2006). *Data analysis using regression and multilevel/hierarchical models*. Cambridge University Press, New York.

Gelman, A., Hwang, J., and Vehtari, A. (2014b). Understanding predictive information criteria for Bayesian models. *Statistics and Computing*, 24(6):997–1016.

Gelman, A. and Shalizi, C. R. (2013). Philosophy and the practice of Bayesian statistics. *British Journal of Mathematical and Statistical Psychology*, 66(1):8–38.

Gelman, A. and Su, Y.-S. (2018). *arm: Data analysis using regression and Multilevel/Hierarchical Models*. R package version 1.10-1.

Ghasemi, A. and Zahediasl, S. (2012). Normality tests for statistical analysis: A guide for non-statisticians. *International Journal of Endocrinology and Metabolism*, 10(2):486.

Goad, H., Guzzo, N. B., and White, L. (2020). *Parsing ambiguous relative clauses in L2 English: Learner sensitivity to prosodic cues*. Studies in Second Language Acquisition. To appear.

Greenland, S., Senn, S. J., Rothman, K. J., Carlin, J. B., Poole, C., Goodman, S. N., and Altman, D. G. (2016). Statistical tests, P values, confidence intervals, and power: A guide to misinterpretations. *European Journal of Epidemiology*, 31(4):337–350.

Gries, S. T. (2013). *Statistics for linguistics with R: A practical introduction*. Walter de Gruyter, Berlin, 2nd edition.

Herschensohn, J. (2013). Age-related effects. In Herschensohn, J. and Young-Scholten, M., editors, *The Cambridge handbook of second language acquisition*, pages 317–337. Cambridge University Press, New York.

Hu, Y. and Plonsky, L. (2020). *Statistical assumptions in L2 research: A systematic review*. Second Language Research.

Jaeger, T. F. (2008). Categorical data analysis: Away from ANOVAs (transformation or not) and towards logit mixed models. *Journal of Memory and Language*, 59(4):434–446.

Kay, M. (2020). *tidybayes: Tidy Data and Geoms for Bayesian Models*. R package version 2.1.1.

Kruschke, J. K. (2015). *Doing Bayesian data analysis: A tutorial with R, JAGS, and Stan*. Academic Press, London, 2nd edition.

Kruschke, J. K. and Liddell, T. M. (2018). Bayesian data analysis for newcomers. *Psychonomic Bulletin & Review*, 25(1):155–177.

Kuznetsova, A., Brockho, P., and Christensen, R. (2017). ImerTest package: Tests in linear mixed effects models. *Journal of Statistical Software, Articles*, 82(13):1–26.

Lee, M. D. and Wagenmakers, E.-J. (2014). *Bayesian cognitive modeling: A practical course*. Cambridge University Press, Cambridge.

Levshina, N. (2015). *How to do linguistics with R: Data exploration and statistical analysis*. John Benjamins Publishing Company, Amsterdam.

Loewen, S. (2012). The role of feedback. In Gass, S. and Mackey, A., editors, *The Routledge handbook of second language acquisition*, pages 24–40. Routledge, New York.

Luke, S. G. (2017). Evaluating significance in linear mixed-effects models in R. *Behavior Research Methods*, 49(4):1494–1502.

Lyster, R. and Saito, K. (2010). Oral feedback in classroom SLA: A meta-analysis. *Studies in Second Language Acquisition*, 32(2):265–302.

Mackey, A. and Gass, S. M. (2016). *Second language research: Methodology and design*. Routledge, New York, 2nd edition.

Marsden, E., Morgan-Short, K., Thompson, S., and Abugaber, D. (2018). Replication in second language research: Narrative and systematic reviews and recommendations for the field. *Language Learning*, 68(2):321–391.

Matuschek, H., Kliegl, R., Vasishth, S., Baayen, H., and Bates, D. (2017). Balancing Type I error and power in linear mixed models. *Journal of Memory and Language*, 94:305–315.

McElreath, R. (2020). *Statistical rethinking: A Bayesian course with examples in R and Stan*. Chapman & Hall/CRC, Boca Raton, 2nd edition.

McGrayne, S. B. (2011). *The theory that would not die: How Bayes' rule cracked the enigma code, hunted down Russian submarines, & emerged triumphant from two centuries of controversy*. Yale University Press, New Haven.

Nakagawa, S. and Schielzeth, H. (2013). A general and simple method for obtaining R2 from generalized linear mixed-effects models. *Methods in Ecology and Evolution*, 4 (2):133–142.

Norouzian, R., de Miranda, M., and Plonsky, L. (2018). The Bayesian revolution in second language research: An applied approach. *Language Learning*, 68(4):1032–1075.

Nuzzo, R. (2014). Scientific method: Statistical errors. *Nature News*, 506(7487):150.

Plonsky, L. (2013). Study quality in SLA: An assessment of designs, analyses, and reporting practices in quantitative L2 research. *Studies in Second Language Acquisition*, 35 (4):655–687.

Plonsky, L. (2014). Study quality in quantitative L2 research (1990–2010): A methodological synthesis and call for reform. *The Modern Language Journal*, 98(1):450–470.

Plonsky, L. (2015). *Advancing quantitative methods in second language research*. Routledge, New York.

R Core Team (2020). *R: A language and environment for statistical computing*. R Foundation for Statistical Computing, Vienna, Austria.

Russell, J. and Spada, N. (2006). The effectiveness of corrective feedback for the acquisition of L2 grammar. *Synthesizing Research on Language Learning and Teaching*, 13:133–164.

Sarton, G. (1957). *The study of the history of science*. Harvard University Press, Cambridge, MA.

Schwartz, B. D. and Sprouse, R. (1996). L2 cognitive states and the full transfer/full access model. *Second Language Research*, 12(1):40–72.

Sonderegger, M., Wagner, M., and Torreira, F. (2018). *Quantitative methods for linguistic data*. Available at http://people.linguistics.mcgill.ca/~morgan/book/index.html.

Stan Development Team (2020). *RStan: The R interface to Stan*. R package version 2.19.3.

Tenenbaum, J. B., Griffiths, T. L., and Kemp, C. (2006). Theory-based Bayesian models of inductive learning and reasoning. *Trends in Cognitive Sciences*, 10(7):309–318.

Venables, W. N. and Ripley, B. D. (2002). *Modern applied statistics with S*. Springer, New York, 4th edition. ISBN 0-387-95457-0.

Watanabe, S. (2010). Asymptotic equivalence of Bayes cross validation and Widely Applicable Information Criterion in singular learning theory. *Journal of Machine Learning Research*, 11(Dec):3571–3594.

Wheelan, C. (2013). *Naked statistics: Stripping the dread from the data*. WW Norton & Company, New York.

White, L. (1989). *Universal grammar and second language acquisition*, volume 1. John Benjamins Publishing, Amsterdam.

Wickham, H. (2016). *ggplot2: Elegant graphics for data analysis*. Springer, London, 2th edition.

Wickham, H. (2017). *tidyverse: Easily Install and Load the 'Tidyverse'*. R package version 1.2.1.

Wickham, H. et al. (2014). Tidy data. *Journal of Statistical Software*, 59(10):1–23.

Wickham, H., François, R., Henry, L., and Müller, K. (2020). *dplyr: A grammar of data manipulation*. R package version 0.8.5.

Wickham, H. and Grolemund, G. (2016). *R for data science: Import, tidy, transform, visualize, and model data*. O'Reilly Media, Inc., Sebastopol, CA.

Wickham, H. and Henry, L. (2019). *tidyr: Tidy Messy Data*. R package version 1.0.0.

Winter, B. (2019). *Statistics for linguists: An introduction using R*. Routledge, New York.

SUBJECT INDEX

%in% operator 26

AIC 153
ANOVA 51

bootstrapping 12, 95
boxplots 75

comments in R 21
correlation test 71
csv files 31
cumulative probabilities 174

data frames 28
DIC 197

effect sizes 9
effective sample size ($\hat{n}_{eff}$) 225
equality operator 26, 41

file organization 32, 64
font families for figures 103
functions 25

Gaussian distribution 50
Gelman-Rubin diagnostic ($\hat{R}$) 225
generalized linear models (GLM) 148
ggplot2 43

Hamiltonian Monte Carlo (HMC) 218
highest density interval (HDI) 216
histograms 68
homoscedasticity 114

indices 25
inequality operator 41
interpreting interactions 128, 165, 184

Likert scale 173
lists 27
local or global specification 72
log-odds 146
logicals 26
logistic curve 145
logit function 145
long tables 66
long-to-wide transformation 42

math operations 22
maximum likelihood 148
mean 26
model comparison 131
multiple linear regression 123

posterior predictive check 228
preambles 36
probabilities, odds, and log-odds 146

Q-Q plots 114

R packages 15
R projects 32
R^2 112–113, 119, 123–125
residual sum of squares 131
residuals 112
RStudio's console 17
RStudio's interface 17
RStudio's script window 17

sampling distribution of the sample
  mean 12
saving your plot 45
scatter plots 70
shrinkage 201
simple linear regression 115
slice notation 29

standard deviation 12, 26
standard error 11, 44

t-test 51
thresholds (τ) 175
tibbles 31, 40
tidyverse 35
trace plot 225
trend lines 71

variables 22
vectors 22

wide-to-long transformation
  36, 66
working directory 19

FUNCTION INDEX

anova() 19
arrange() 40

brm() 224

c() 24
citation() 59
clm() 177
clmm() 194
complete() 98
coord_flip() 102

fct_reorder() 77
filter() 40

geom_bar() 45, 83
geom_boxplot() 45, 77
geom_histogram() 69
geom_jitter() 77
geom_point() 45, 71
geom_rug() 71
geom_text() 188
geom_vline() 103
glimpse() 40
glm() 148
glmer() 194
group_by() 40

ifelse() 149
install.packages() 36

lm() 116

mutate() 40

nest_by() 255

predict() 129

relevel() 157
remove.packages() 59
rescale() 135
rnorm() 50

sample() 50
scale_color_manual() 73
scale_fill_brewer() 96
scale_y_continuous() 102
select() 40
separate() 67
set.seed() 59
stat_smooth() 71
stat_summary() 79
summarize() 40

theme() 102

update.packages() 55

Printed in the United States
by Baker & Taylor Publisher Services